SSSP

Springer
Series in
Social
Psychology

Springer Series in Social Psychology

Attention and Self-Regulation:
A Control-Theory Approach to Human Behavior
Charles S. Carver/Michael F. Scheier

Gender and Nonverbal Behavior
Clara Mayo/Nancy M. Henley (Editors)

Personality, Roles, and Social Behavior
William Ickes/Eric S. Knowles (Editors)

Toward Transformation in Social Knowledge
Kenneth J. Gergen

The Ethics of Social Research:
Surveys and Experiments
Joan E. Sieber (Editor)

The Ethics of Social Research:
Fieldwork, Regulation, and Publication
Joan E. Sieber (Editor)

Anger and Aggression:
An Essay on Emotion
James R. Averill

The Social Psychology of Creativity
Teresa M. Amabile

Sports Violence
Jeffrey H. Goldstein (Editor)

Nonverbal Behavior: A Functional Perspective
Miles L. Patterson

Basic Group Processes
Paul B. Paulus (Editor)

Attitudinal Judgment
J. Richard Eiser (Editor)

SSSP

Social Psychology of Aggression

From Individual Behavior to Social Interaction

Edited by
Amélie Mummendey

Springer-Verlag
Berlin Heidelberg New York Tokyo 1984

Professor Dr. Amélie Mummendey

Westfälische Wilhelms-Universität, Psychologisches Institut, Schlaunstrasse 2, D-4400 Münster/W.

H M
291
. S 5886
1984

With 17 Figures and 18 Tables

ISBN 3-540-12443-8 Springer-Verlag Berlin Heidelberg New York Tokyo
ISBN 0-387-12443-8 Springer-Verlag New York Heidelberg Berlin Tokyo

Library of Congress Cataloging in Publication Data
Main entry under title: Social psychology of aggression. (Springer series in social psychology). Based on papers presented at a conference held Oct. 1983 at the Zentrum für Interdisziplinäre Forschung at the University of Bielefeld. Bibliography: p. Includes indexes. 1. Aggressiveness (Psychology)-Congresses. 2. Social psychology-Congresses. I. Mummendey, Amélie, 1944–. II. Series. HM291.S5886 1984 302.5'4 83-27096

© Springer-Verlag Berlin Heidelberg 1984
Printed in Germany

Typesetting and bookbinding: G. Appl, Wemding. Printing: aprinta, Wemding
2126/3140-543210

uly 84

Foreword

Dollard, Doob, Miller, and Mowrer formulated their frustration-aggression hypothesis more than forty years ago. Since then the progress in theory of and research on aggression has been very slow. Today we know that there are severe limitations to their hypothesis. The development of alternative approaches has been restricted by the neglect of sociopsychological concepts. Until a few years ago, social psychology was at the back door of aggression research, and even this superficial acquaintance contained too many cognitive ideas to suit many of the influential heroes of the mainstream of research.

There are many reasons for the decline of the old paradigms in aggression research, among them the failure to extrapolate from the results of artificial experiments to the realities of our time. This book goes much deeper than other texts in the area; it is also a fresh beginning. It endeavors to reformulate the more traditional topics and strongly emphasizes the social framework of aggression. Accordingly, hostile actions must be explained from a sociopsychological perspective. It has remained for Amélie Mummendey to show the way in which European and American research can be effectively integrated in a comprehensive reader on aggression.

Until recently I considered the topic of aggression to be about the most tedious in social psychology, but in this book the subject takes on exciting new life because of the emphasis on social interaction and the integration of cognitive and motivational concepts. According to this analysis, attributions, moral judgments, subjective definitions of the social situation, antinormative behavior, acceptability of justifications and excuses, considerations of justice, and social interdependence are the key concepts in an explanation. In addition, individual differences in aggressiveness are taken into account. The authors examine the contribution social psychology might make to an analysis of aggression. The contributions provide excellent leads to the meaning and the role of illegitimate and antinormative aspects in the problem of aggression.

Marburg, March 1984   Hans Werner Bierhoff

Preface

Nobody would provoke a big argument by making the statement that *aggression* is one of the most prominent research topics in the behavioral and social sciences. The importance of the topic led to the establishment of the International Society for Research on Aggression (ISRA), which is an interdisciplinary organization oriented to various natural sciences like biology, physiology, pharmacology, and psychology. Social sciences like sociology, political science, and social philosophy deal with this topic within the context of research on all kinds of conflicts and conflict resolution. The aim of this volume is to provide a perspective on social psychology as something between the behavioral and social sciences, concerned with aggression as a particular kind of social interaction – be it between two or more individuals or between two or more groups or larger social units. The fundamental characteristics of this interaction *transcend* individual behavior as described in strictly observational terms, i.e., they are related to constructions of social reality like norms and norm violations, intention and responsibility or accident, and injustice or legitimation of criticism. The various disciplines in the social sciences deal with these different aspects, each in their own way; their contributions could be fruitful for the further elaboration of social psychological concepts of aggressive interaction.

The present volume contains contributions from different disciplines like psychology, social psychology, sociology, and social philosophy, which represent work done in various European countries as well as in Canada and the United States of America. It is based on discussions at a conference held at the Zentrum für interdisziplinäre Forschung (ZIF) at the University of Bielefeld in October 1982. I gratefully acknowledge the generous support provided by the ZIF, which made it possible for participants from various countries on each side of the Atlantic to attend the meeting.

Due to the very efficient organization of the meeting and the very friendly cooperation of Springer-Verlag, it turned out to be a real pleasure and an easy task to realize the plan of editing this book. I want to thank, therefore, especially Joan Goldstein, Stephanie Benko, and Thomas Thiekötter. A particular acknowledgement goes to Michael Wilson; he had the hard task of transforming the contributions by the non-English-speaking authors into comprehensible English versions. If the aim has not been completely reached, this should not be attributed to him. Last but not least, I thank Sibylle Claßen for doing a large part of the typing and Birgitta Sticher for accurately assisting in the preparation of the index.

Münster, March 1984   Amélie Mummendey

Contents

Introduction
    Amélie Mummendey 1

1. A Social Psychological Interpretation of Human Aggression
    James T. Tedeschi 5

Current Concepts of Aggression. 5
The Study of Threats and Punishments 10
Laboratory Research on Aggression. 17
Conclusion 19

2. Motivation Theory of Aggression and Its Relation to Social
    Psychological Approaches
    Hans-Joachim Kornadt 21

Motivation Theory of Aggression 23
Relationship Between Motivation Theory and Social Psychological
Approaches. 27
Essentials of the Motivation Theory. 28
Conclusion 30

3. Individual Differences and Aggressive Interactions
    Horst Zumkley. 33

Social Psychological Perspective: "Aggressive" Is Always the Other Person . 33
"Aggressive" Is Always the Other Person?: Some Research Gaps 34
Retaliation Norm in Aggressive Interactions and Individual Differences . . . 35

Causal Ambiguity in Aggressive Interactions and Individual Differences . . .   41
Concluding Remarks   47

4. *Aggression as Discourse*
    Kenneth J. Gergen   51

Aggression as Linguistically Grounded   54
The Structural Unpacking of Aggression Discourse   57
On the Negotiation of Aggression   64
Summary   67

5. *Aggression: From Act to Interaction*
    Amélie Mummendey, Volker Linneweber, and Gabi Löschper   69

Mutually Interpreting Behavior in Aggressive Interactions   76
The Context of Aggressive Interactions: Taxonomy of Social Situations . . .   84
The Actor-Victim Divergence in Aggressive Interactions   92
Conceptions About the Progress of Aggressive Interactions   98
Conclusion 100

6. *Patterns of Aggressive Social Interaction*
    Richard B. Felson 107

Theoretical Approaches 108
A Description of Aggressive Incidents 113
Discussion 123

7. *Frustration, Aggression, and the Sense of Justice*
    Jorge da Gloria 127

Reactions to Failure, Obstruction, and Attack 132
Coping with Failure 134
Coping with Obstruction and Attack 135

8. *The Relations Among Attribution, Moral Evaluation, Anger,
    and Aggression in Children and Adults*
    Brendan G. Rule and Tamara J. Ferguson 143

The Role of Norms and Causality in Blame, Anger, and Aggression 143
Biasing Conditions 148
Conclusions 150
General Issues 151

9. Social Justice and the Legitimation of Aggressive Behavior
  Dieter Birnbacher 157

Three Uses of the Concept "Aggression" 158
Principles of Social Justice and Their Role in Aggressive Interaction 159
The Utilitarian Way Out 168

Author Index 171

Subject Index 174

List of Contributors

Dr. Dieter Birnbacher
Arbeitsgruppe Umwelt Gesellschaft Energie (A.U.G.E.), Universität Essen
FB1, Universitätsstrasse 2, D-4300 Essen 1

Dr. Jorge da Gloria
Université Paris VII, UER de Sciences Sociales, Laboratoire de Psychologie
Sociale, 18, rue de la Sorbonne, F-75230 Paris Cedex 05

Dr. Richard B. Felson
Department of Sociology, State University of New York at Albany, 1400 Washington Avenue, Albany, NY 12222, USA

Dr. Tamara J. Ferguson
Vakgroep Ontwikkelingspsychologie, Katholieke Universiteit, Montessorilaan 3,
N-6500 HE Nijmegen

Prof. Dr. Kenneth J. Gergen
Department of Psychology, Swarthmore College, Swarthmore, PA 19081, USA

Prof. Dr. Hans-Joachim Kornadt
Fachbereich 6 der Universität des Saarlandes, Fachrichtung Allgemeine Erziehungswissenschaft, Dienstgebäude: Bau 11, 2. OG, D-6600 Saarbrücken

Dr. Volker Linneweber
Westfälische Wilhelms-Universität, Psychologisches Institut, Schlaunstrasse 2,
D-4400 Münster/W.

Dr. Gabi Löschper
Westfälische Wilhelms-Universität, Psychologisches Institut, Schlaunstrasse 2,
D-4400 Münster/W.

Prof. Dr. Amélie Mummendey
Westfälische Wilhelms-Universität, Psychologisches Institut, Schlaunstrasse 2, D-4400 Münster/W.

Prof. Dr. Brendan G. Rule
Department of Psychology, Edmonton, Alberta T6G, 2E9, Canada

Prof. Dr. James T. Tedeschi
Department of Psychology, State University of New York at Albany, 1400 Washington Avenue, Albany, NY 12222, USA

Dr. Horst Zumkley
Fachbereich 6 der Universität des Saarlandes, Fachrichtung Allgemeine Erziehungswissenschaft, Dienstgebäude: Bau 11, 2. OG, D-6600 Saarbrücken

Introduction

Amélie Mummendey

Among the large variety of possible conditions and situations which can prevail when a book on a particular topic is being prepared, two extreme conditions can be imagined. First, the topic is completely new, and nothing has been written about it; second, the topic is not new at all; many people have been working on it for years, and numerous books and publications are available and still appearing. At each of these extremes, some kind of explanation for working and publishing on that particular topic seems to be demanded by the potential recipients. In the first case, therefore, the explanation will refer to reasons for dealing with that particular topic at all, reasons why these questions are raised and thought about. In the second case, explanations will refer to reasons for *once again* presenting a publication after so many have already appeared – justifications for the author's opinion that particularly this publication should be useful in addition to so many others, and which make it different from others and therefore necessary.

The present book apparently belongs to the second type. The explanation justifying another publication on aggression, therefore, has to focus on those features of the concept of this book which make it different from the others. To judge whether it is necessary or useful, of course, is up to the reader.

Consistent with the above remarks, the explanation and justification for publishing this book will not start once again by mentioning the extraordinary social relevance of aggression as a research topic in various areas of psychology. To describe the aim of this book and how the different chapters are related to it, I want to start with an example, a tragedy which everybody has witnessed in some way and which happened just when I started writing this introduction: In the very early morning of September I, 1983, an aircraft of the Korean airlines was shot down by a Soviet fighter; 269 people were killed, among them 19 children.

In uni son, the West German media and – as far as I can tell – those of other Western countries, too, *immediately* spread a reaction from officials which vicariously expressed the attitude of large parts of the general population: Emotions were expressed like indignation, rage, anger, dread, or horror about the *action*. The action was unanimously interpreted as calculated deliberate homicide or, referring to the Soviet government as subornation to homicide. Evaluations referred to the incomparable brutality, cruelty, and terroristic nature of the action. The first explanatory parts of those reactions were restricted to the statement on the one hand, that this action was incomprehensible, that NATO pilots would never get such an order, and that if they did get it they would refuse to obey to it. On the other hand, according to the initial reactions the act was not astonishing because the Soviet regime represents the "imperium of the evil" and is capable of everything. Beyond any doubt or mitigation, the Soviet Union was identified as the instigator, agent of crime, and aggressor. The action was declared unjustifiable and there was a self-evident claim for the necessity of punishment to serve justice.

As I said before, this was the unequivocal interpretation and reaction during the first days after the tragedy. What are the conditions which produced such an unequivocal view of the event so quickly? Which selection and configuration of information led to this socially shared and firm conviction that the Soviet government is to be condemned and retaliatory measures must be taken.

Apparently the situation was composed of the following facets: A terrible evil was performed on a large number of people; although they were completely peaceful and harmless citizens, they were killed by a Soviet pilot. The pilot performed his action with full consciousness, completely in agreement with his headquarters and after pursuing the Korean aircraft for more than two hours. This means that the Soviet government is perceived as the attacker or as the initiator of the action, the attack being carried out with intent, causing incredible destruction and harm, and being responsible for the deaths of all the people – a violation of a number of fundamental norms and rules.

And apparently at the same time, the following facets were *omitted* (or not considered until during later periods of a more differentiated analysis): What could have been the reasons for the Soviet officials to shoot down this civilian aircraft? What could have been their interpretation of the situation? Are there circumstances conceivable which in the given situation would make the action at least comprehensible? What happened before the critical action? What had the victim done before the critical action was performed? Did the Soviets, from their perspective, really shoot down a "merely" civilian aircraft, or was it also a spy plane which after a number of warnings was attacked completely in accordance with international conventions, making its destruction only a logical consequence and the United States responsible for the deaths? Did the Soviets know that they would kill such a large number of innocent people?

If we look at the accounts provided by Soviet officials, a selection of information is used which has nearly no overlap with that used in the Western interpretation.

I do not know what the facts really are. But a careful analysis of this tragedy certainly would bring to light a complex and many-layered picture.

Unfortunately the Manichaean identification of a person or personlike social unit as evil, making her or him fully responsible for catastrophic events and fighting against her or him in holy rage, is too simple and – as I would dare to say – never adequate in the sense of a more careful view of the events. It would be too simple a method to look for answers to the questions of why people are killed by other people, or why people offend, maltreat, torment, and torture other people. And what is of even greater importance, such a simple model would lead to inferences and conclusions which certainly would miss the goal of avoiding such events and actions.

The social and psychological problem of aggression comprehends more than an individual or individualized action. As can be seen from the given example as well as from innumerous other examples of interpersonal or intergroup aggression, the critical action or event is embedded in a sequence of events happening between individuals or social units. Each participant is oriented by his/her own *partial analysis* of antecedents and consequences, which each of them takes for the whole story and which results in more or less explicit controversies about the interpretation and evaluation of the critical event. The strong convictions of the various parties with their respective interpretations and beliefs provide a favorable basis for further actions (like punishment!) suited not to stop but on the contrary to continue the exchange of wrongdoing.

The contributions to the present book deal with various factors which are of interest in such a broader view of the phenomenon of human aggression. Although there are considerable differences between the various contributions, the following global orientation is common to all of them: The genuinely social characteristics of the phenomenon "aggression" are stressed, and there is a consensus about conceptualizing it basically as a kind of social interaction. This means that, beyond the mere behavior, the exchange between two individuals or social units is looked upon; therefore the critical event is seen as part of a sequence of actions and reactions. Interpretations of the situation and evaluations of the critical action come to be of interest. There are questions about the conditions which in the view of the respective participant permit the construct "aggression" to be formed from the occurrence of harmful events. Thus, conceptions of rules, norms, and morality move to the center of interest in this context.

Tedeschi points out the limitations of current concepts of aggression in psychology. He proposes instead an approach toward the problem consisting of first, systematizing its behavioral part in a strictly behavior-oriented terminology and taxonomy of threats and punishments as different kinds of interpersonal influence, and second looking for systematic regularities in interpersonal judgments and evaluations of the behavior in question. Kornadt and Zumkley stress the importance of integrating the analysis of individual differences into the sociopsychological concept of aggressive interaction. Kornadt discusses his theory of aggression motivation in this context, e.g., its relations to social behavior. Coming from the same background, Zumkley discusses a number of recent interrelated experimental investigations which provide impressive evidence of cross-situational consistency for individual differences in aggressiveness and their influence upon interpreting and explaining critical events, e. g., the tendency to attribute hostile motivations. Thus information and thoughts on further research into the kind and function of individual differences in aggressive social interactions become evident.

The process of generating the social fact "aggression" as part of the individual's construction of social reality is analyzed by Gergen. He proposes a linguistically grounded approach to aggression as the product of language conventions, forming the interpersonal or social "discourse." By the method of horizontal and vertical "structural unpacking," the "structural nucleus" of the term and the neighboring nuclei which embed the central criteria of the structural nucleus are revealed.

Mummendey, Linneweber, and Löschper present an approach and report some studies which focus on characteristic features of aggressive interactions as a sequence of exchanges between two individuals or social units typically related to each other as actor and recipient within a conflict situation. Felson is also concerned with the interaction sequence and especially concentrates on subjective perceptions of antecedents for one's own aggressive actions within a sequence of events. Aggressive actors' accounts of various aspects of the critical sequence of interactions are analyzed and discussed with respect to different theoretical conceptions of aggression, such as impression management, coercive power, and punishment.

Da Gloria presents a detailed analysis of the frustration – aggression hypothesis and experimental research related to it. As to the analysis of aggression, he seeks to distinguish two kinds of frustration from perceived disconfirmation of predicted outcomes, i. e., "failure" and "obstruction and attacks." Aggressive interactions occur in response to the second kind and are thus conceptualized as the (mutual) restoration of self-evaluation with respect to the ability of an actor to control the behavior of another to the extent previously expected. This expectation is oriented on normative prescriptions and conventions, for example of justice. Coming from a somewhat different background, Rule and Ferguson deal with a related question. They analyze the effect of different normative beliefs in relation to different kinds of causal attributions on the evaluation of harmful actions and reactions, e. g., sanctions. In this context and among other items, the conditions *biasing* the process of the evaluation of a harmful act as aggressive and the development of the desire to punish or retaliate are discussed, demonstrating very clearly the highly complex social conditions which are essential for the perception of an action as aggressive and the selection of an appropriate reaction to it.

Finally, Birnbacher, from a philosophical point of view, considers social justice as a philosophical concept, its relation to aggression or violence, and its justification. He especially stresses the relevance of the concept of procedural justice as compared with that of substantive justice with regard to the aim of minimizing aggressive interactions; although, at the same time, the existence of opposed conceptions of substantive justice or values of two opponent individuals or social units may be conceived to be inevitable, agreement on procedural justice may be a way of minimizing aggression and violence.

Chapter 1

A Social Psychological Interpretation of Human Aggression

James T. Tedeschi

Near the end of the most bloody century in the history of the human species it is noteworthy how little progress has been made in the study of aggression. Our knowledge has advanced little further than a theory proposed by Freud (1920) more than half a century ago. Although Freud's frustration-aggression theory was recast into the language of behavioristic learning theory (Dollard, Doob, Miller, Mowrer, and Sears, 1939), the basic principles are little changed from the original formulation. Alternative psychological theories of instrumental conditioning, cue-frustration and aggression, and modeling have not provided great insights into why people assault, rape, and kill one another. The view to be presented here is that the essential problem with psychological approaches to studying human aggression is an inadequate conceptualization of the phenomena to be explained. Current conceptions of aggression will be examined and an alternative interpretation will be proposed.

Current Concepts of Aggression

Behavioristic Definition of Aggression

Anyone attempting to define aggression as a set of denotative striate muscle responses must carefully designate which behaviours are to be included in the set and which are to be excluded. That is, we must be able to discrimate between behaviors to be characterized as aggressive and those that are not. This task is not very difficult when considering nonhuman organisms because they emit rather sterotyped behaviors that have predatory and agonistic functions. Consider a house cat. The predatory pattern of stealth, patience, and attack, pounce, and bite is recognizable by anyone who is familiar with cats. The agonistic pattern is also easily described: hair reaised on back, hunched back, claws unsheathed, and hissing and screaming sounds. While different striate muscle reactions may be involved, adequate descriptions can also be given for the aggressive actions of a bull, a stickleback fish, or a gibbon.

Unfortunately, no one has been successful in designating a taxonomy of human aggressive responses. Given the impossibility of such a task because of the great variety of human behavior, there has been a shift toward trying to find out what characterizes a set of aggressive actions. Figure 1 indicates a set of responses, R_1, R_2, ..., Rn, which exhausts the behaviors we wish to designate as aggressive. The question is: what is similar about these behaviors which makes them different from nonaggressive behaviors?

R_1
R_2
R_3
I
I
I
R_N

Fig. 1.1. The set of responses to be characterized as aggression

Behaviorists, believing they were describing behavior, used the criterion of harm done. Any behavior that had the effect of harming another living organism was an instance of aggression. This criterion served to discriminate between aggressive and nonaggressive behaviors. Before long, various problems with this definition began to be recognized even by the behaviorists themselves (e.g., Buss, 1961). The criterion tended to include in the set of aggressive behaviors some that seemed intuitively not to belong, and it did not include other behaviors which everyone agreed ought to be included.

The behavioristic definition of aggression includes accicental harm-doing. Actually, this is not surprising if one considers that the definition was developed to designate the behavior of subhuman animals. After all, lions and wolves seldom have accidents. The frequency or incidence of accidents appears to be directly related to the use of tools, particularly of machines, which are employed to mediate the goals of people. Humans have learned to differentiate between actions that are performed by choice ("free will") and those which are a result of forces few people could be expected to resist. People do not attribute free will and choice among consequences to subhuman animals. If we include inadvertency, mistake, and accident together in one category, it constitutes a significant exception to the behavioristic definition of aggression. Doing harm by itself does not discriminate between aggressive and nonaggressive behavior. In any case, behaviorists do exclude accidents from their definition but have not proposed an adequate criterion to replace the harm-doing effects of behavior.

There are also behaviors excluded by the harm-doing criterion which everyone would agree should be included in the set of behaviors to be labeled as aggression. These are actions which are believed by observers to have the intent to do harm but fail to produce it. Consider the case of a man placing a bomb in an automobile and wiring it to the ignition. Assume that because of a mistake the bomb does not work and suppose the driver starts the auto and drives off unaware of the presence of the bomb. Despite the fact that no physical or psychological harm was done, most people would want to include the action of the would-be bomber in the set of behaviors to be labeled as aggression. By the harm-doing criterion, however, it would not be considered by the behaviorists to be aggression. We must conclude that the behavioristic definition is inadequate on at least two grounds: it includes too much and too little.

Attributional Definition of Aggression

Recognition of the twin problems associated with the behavioristic definition of. aggression led to a new criterion to discriminate between aggressive and nonaggressive behavior. Any behavior that has the intent to do harm is to be considered aggression. This criterion allows one to distinguish between accidents and aggressive behavior, since the former is unintentional (as are inadvertency and mistake). Also, this criterion allows us to include behaviors that do not do any harm but which are believed to be emitted with the expectation by the actor that harm would be done. On intuitive grounds the attributional definition appears superior to and remedies the problems of the behavioristic one.

On closer examination, however, the attributional definition appears to have insuperable problems of its own. Perhaps the most unrecognized problem is that no one has provided an adequate scientific definition of "intention" nor have any criteria been proposed for establishing when inferences of intent should be made.[1] This failure has led to the interpretation of fantasy as aggression (Feshbach, 1955). One cannot distinguish among the mere thought of aggression, a wish to do harm, contingency planning, and a real intent to do harm.

This is not the place to get into a long discussion about the concept of intention. Anscomb (1963) revealed the logical fallacy of believing it to be an internal cause of behavior – an instance of what Ryle (1949) has called a category mistake. The best we can say about intention is that it involves an inference on the part of the observer (even if he is also the actor) regarding what the actor could have said about the effects of his behavior before undertaking the action. That is, an intentional action is one having effects the actor could have articulated, prior to the action. There is nothing inherent in an action that requires such an inference. Intentions are not actions nor are they observable, but are aspects of the attribution process on the part of observers. The implication for those who would use the criterion of intent to identify aggressive actions is that each scientific observer will have to make his or her own inferences about intent. But then all pretense of having a denotative definition of a set of discriminable and directly observable responses sharing a single criterion should be given up. The attributional definition is purely a matter of the phenomenology of the scientific (or lay) observer.

There is a second important problem with the attributional definition of ag-

---

1 We must differentiate between a naive psychology of person attributions (e.g., Heider, 1957) and an adequate scientific construct of intention. The latter presumably would contribute to a social psychological theory of action.

gression. The definition includes behaviors that intuition suggests should not be considered as aggression. Consider the following example. I come to your house in the middle of the night and chop down your front door with my trusty ax. It is not possible to consider such an action as an accident. Physical harm was done to your home, perhaps making the house cold and requiring the expenditure of money for repairs. That is, harm was clearly done to you. This seems clearly an act of aggression. But what if I had on a fireman's uniform, had come on a fire truck with other men, and believed there was a fire in your house? Well, in that instance, though you might not be happy about the loss of your door, you would probably not see the action as an instance of aggression. In each example, I had the immediate intent to break down the door. The difference in the two descriptions is the presumed remote goal of the action. In the first example, no remote goal is stated. In the second example, the actor is believed to have sufficient justification for the action. The remote goal associated with axing the door is to put out the fire and possibly save lives. This remote goal is implicitly communicated through the fireman's apparel. Thus, axing the door, a harm-doing action, may or may not be perceived as aggression, even though it fulfills the criteria of the attributional definition. It surely was not an accident, and attribution theorists would probably consider the action to be intended. The critical factor appears to be one of justification.

What is justified is not a scientific question, though of course we can examine why and how people make such judgments. There are multiple values and norms serving as justifications for actions (see Tedeschi & Riess, 1981, for a discussion of some of these). People, including social psychologists, differ in their values and in their cultural backgrounds. The acceptable justifications for behavior are therefore variable and a matter of subjective processes. If some criterion of what makes an intentional harm-doing behavior justifiable had to be adopted prior to an identification of an aggressive behavior, our task would be impossible. Thus, the two basic problems with the attributional definitions are associated with the subjective judgments of intent and justification, which must be standardized before a satisfactory, value-free scientific definition of what has heretofore been considered an observational concept of aggression could be achieved.

Aggression as Release of Energy

Dollard et al. (1939) identified a response as belonging to the set ot be considered aggression if it released aggressive energy. The frustration-aggression theory proposed that frustration, a condition where an ongoing goal response is blocked or thwarted, leads to a build-up of aggressive energy within the organism. This energy is noxious and must be released by the organism in the form of aggressive behavior. Any response that releases this aggressive energy is an instance of aggression. The circularity of this definition is obvious. Without an independent ability to observe and measure the presence, accumulation, and release of aggressive energy, it is impossible to identify responses as aggression.

Aggression as a Skill

Bandura (1973) views aggression as behavior patterns that are learned, largely through reinforcements and modeling, which can be intentionally directed towards harming others. There are three aspects to learning aggression: the acquisition, maintenance, and performance of the skills.

Consider the typical modeling study carried out by Bandura and his associates. An adult model is placed in a room with a set of props, usually including a large vinyl Bobo clown that bounces down to the floor and back up when pushed, kicked, or punched; a rubber mallet; rubber-tipped darts and gun; and so on. According to prior planning, the adult punches and kicks Bobo, often with accompanying verbal statements, such as, "Take that you bad Bobo," shoots Bobo with the dart gun, and hits it with the mallet. After some minutes of this kind of activity, another adult (the experimenter) praises the actor and gives her a reward (e.g., a candy bar). Children who observe a rewarded adult model tend to imitate the behavior and engage in more of these same behaviors than children who observe a model who is unrewarded or who observe no model at all.

The imitative behavior of the children is called aggression. But in what sense is it aggressive? The behavior clearly does not meet the behavioristic criterion of aggression, since no harm is typically done to Bobo. It seems implausible to invoke the attributional definition because there is little reason to believe the children want to damage Bobo or even that they believe they could after witnessing an adult fail to do so. Furthermore, rewarding the model probably serves to justify the conduct of the children. Thus, the behavior appears to lack intent to do harm, and, even if intent were present, the behavior appears justifiable under the circumstances.

Bandura considers the imitative behavior of the children as instances of aggression because it includes skills, such as punching, kicking wielding and directing a blow with a blunt instrument, and shooting a gun, which, if directed at a live target, could do harm. Skills must be acquired or learned, but the individual does not need to do harm to learn them. For example, a person can learn to become a marksman with firearms at a firing range. Skills must be maintained or they become rusty or inadequate for producing intended effects. If they are used to produce effects, such intended behaviors are referred to as performance. For Bandura it is important to distinguish between learning and performance. Unfortunately, the distinction requires the identification of intent.

Even if there were not a problem with Bandura's acceptance of a criterion of intention, it would be impossible to discriminate between aggressive and non-aggressive responses. Consider the multitude of ways in which one person can harm another. What skills are involved? A person sitting in an underground silo can launch a missile with a push of a button and send it to destroy a city full of people. Learning to manipulate and control the finger probably occurred when the person was a baby in a crib; it is a simple skill but not one we have always had. Recently, a woman in Albany, New York was convicted of murder in a jury trial because she hired someone to kill her husband. If she could not talk (or communicate), she could not have conspired to carry out the murder. Learning to talk, then, must be considered a skill that can be used later on to intentionally harm others. Suppose I am walking along a cliff with someone I intensely dislike and I thrust my hip into him and topple him off the side. The action could not have been taken if I had not learned how to walk; hence, walking must be considered an aggressive behavior because it can be used to harm others. With enough examples it becomes clear that almost any human behavior can be used to harm someone else. Thus, if we accept Bandura's definition of aggressive behaviors as acquired skills that can be used to intentionally harm others, we cannot discriminate between aggressive and nonaggressive behaviors. All or almost all behaviors are aggressive. Such an all-inclusive definition is obviously not a scientifically useful one.

Conclusion

Every extant definition of aggression ends with a subjective judgment by the social scientist about questions of causality (accident or not an accident) and about the intentions and justifications of a given action. This is clearly an unsatisfactory way to proceed and may account for part of the problem with the scientific study of human aggression. A fuzzy conceptualization of what it is one is studying will not often lead to clear theoretical explanations of the phenomena of interest. In an attempt to provide an alternative conceptualization of the area to be investigated, Tedeschi and colleagues (Tedeschi, Smith, & Brown, 1974; Tedeschi, Gaes, & Rivera, 1977; Tedeschi, Melburg, & Rosenfeld, 1961; Tedeschi & Melburg, 1983; Tedeschi, 1983) have proposed setting aside problems regarding the labeling of behavior as aggression, attributions of intention, and acceptability of justifications as questions to be resolved by investigators of person perception, and they have offered an alternative set of definitions to identify the phenomena that aggression theorists purportedly are interested in studying.

The Study of Threats and Punishments

The present approach does not try to distinguish between aggressive and non-aggressive behaviors or between intentional and accidental responses. Nor is it concerned, at least on the level of identifying observational constructs, with problems of legitimation and justification. It is assumed that all of these questions will arise at a later point, but first the events to be studied must be identified. The set of events of interest includes all kinds of threats and punishments.

Types and Functions of Threats

There are of course different kinds of threats, but each is a communication from one person to another. A communication is a part of the physical environment, although it of course requires social decoding. That is, the communication is written on paper, is carried through the air by reverberations or as radio signals, and so on. These communications can be considered apart from the person who encoded and transmitted them. It is true that people can perceive threats that do not exist or detect them in communications when they were not intended. But these are questions of person perception and not germane to the present endeavor, which is to provide observational elements for the development of social psychological theory and which would eventually enlighten us about why people rob, rape, assault, and kill one another (among other things).

What kinds of characteristics identify threats as being different from nonthreats? Threats are communications by a source indicating that he will punish a designated target in the future.[2] Contingent threats ask for some sort of compliance from the target, and compliance is presumed to influence the source not to punish the target. Noncontingent threats do not ask for compliance but merely inform the target that the source proposes to punish him. Explicit noncontingent and contingent threats are easy to identify, and a computer with a proper dictionary of terms undertaking content analyses of communications can do so (Holsti, 1968). Tacit or implicit threats are much more difficult to identify, but the task of analyzing language or rules of gestural communication has yet to be undertaken. Studies in other areas provide optimism about the outcome of such analytic work (e.g., Birdwhistle, 1966). In the meantime research on explicit threats can be carried out.

The most obvious use for threats is to compel compliance from a target. The threat may be perceived by the source as having a defensive and deterrent purpose, while the target may perceive it as having an offensive and compellent intent. An actor may attempt by using threats to establish an identity as tough, resistant to intimidation, giving strong allegiance to some person, group institution, or cause. That is, the actor's primary purpose may not be to deter or compel but rather to use intimidating tactics as an impression management strategy (see Jones & Pittman, 1982).

Types of Punishments

Psychologists tend to use Thorndike's (1905) empirical law of effect as a basis for defining punishment: it is something the organism avoids or escapes and does nothing to obtain. These reactions of the organism provide the grounds for identifying a stimulus event as punishment. However, we would limit the range

---

2 A question might arise about threats made when the source is only kidding. From the present viewpoint the communication would still be counted as a threat, but there are circumstances that lead a target to disbelieve it.

of punishments to those that are caused, or at least partially caused by the intervention of human agents. There are enough commonalities in reactions to identify what is punishing for most people within a human culture. Tedeschi (1970) distinguished among four types of punishments: noxious stimulation, deprivation of resources, deprivation of expected gains, and social punishments.

Noxius stimulation refers to physical events that either do tissue damage, and hence have a detrimental biological impact on the organism, and/or produce negative reactions from the organism. The person may not be aware of exposure to radiation, but it may have detrimental biological consequences for him. The placing of a chemical substance on the tongue may result in a bitter taste. Shooting, stabbing, or punching a person would constitute noxious stimulation, since in each case either biological damage or displeasure (or both) occurs.[3]

Deprivation of resources removes, destroys, or takes away from a person something that he possesses and which satisfies needs, promotes security and protection, or enhances power. The most obvious example is a robbery, where the person has his money taken away from him by someone who uses a contingent threat to punish him with noxious stimulation if he does not yield his wallet. A judge may fine a person and thus legally take away his money. The person may have many other physical possessions, such as an automobile, a house, an insurance policy, a bank account, clothing, and so on. The removal of any of these possessions constitutes the form of punishment labeled deprivation of resources. It is assumed that the person experiences dissatisfaction and either resists or desires to resist their removal.

Deprivation of expected gains is a third category of punishments and refers to a situation in which a person expects to achieve a goal, gain resources or some other reinforcement, or remove or avoid an aversive state. Before the desired outcome can be attained, someone intervenes and removes the desired goal. This form of punishment makes salient an aspect of all punishments – a person (or a group of people) must be identified as the cause of the outcome. Punishments are not natural revents, like bolts of lightning (Heider, 1951).

A target who experiences deprivation of expected gains may become very angry. The violation of expectations is apt to be seen as unjust and unfair, and may produce some punitive reaction on the part of the target. Consider the racial integration of public facilities and schools and the extensive civil rights legislation in the United States in the 1950s and 1960s. It was only after this considerable improvement in the legal status of blacks that the violence erupted in the major cities of the country. Why was this the case? The answer is probably very complicated, but one factor was certainly the anger generated by the failure to achieve any improvement in everyday living that blacks had been led to expect as a result of liberal reform. Deprivation of expected gains produces volatile reactions in the target and may be one of the more important ingredients in generating revolutions (Brinton, 1952).

It is interesting to note that deprivation of expected gains appears similar to

---

3 A physician giving a patient an injection is delivering noxious stimulation, but the action is justified by the good it allegedly will do.

the definition of frustration by Dollard et al. (1939). An important difference, however, is that frustration is defined in such as way as to include both human and nonhuman sources of interference in gaining goals.

Social punishments include name-calling; altercasting into negative identities; providing negative evaluations of the target person; exclusion or excommunication from various groups, occupations, and institutions; and degradation ceremonies. The thrust of these actions is to cast the target person into a negative identity, perhaps in front of important audiences. Many arguments that start over a specific problem, such as what television program to watch or whose turn it is to wash the dishes, quickly degenerate into attacks against the target's identity.

While a particular instance of punishment may be predominantly of one type, there are obvious overlaps in the categories. Incarceration, for example, probably involves all four forms of punishment. The person may experience noxious stimulation at the hands of other prisoners. He cannot have access to the resources he may have had outside of prison and so is deprived of them. Whatever goals he was striving to attain prior to incarceration are removed and hence he experiences deprivation of expected gains. And, of course, he has acquired the stigma associated with identities of legal felon and convict, a form of social punishment.

Functions of Punishment

Punishments are often used in association with threats. Once a threat is communicated, the source's reputation is on the line. If the threat is noncontingent or if a target fails to comply to a contingent threat, failure to carry through with the punishment lowers the credibility of the source. If a person establishes a reputation for having low credibility, future threats are not apt to inspire fear or compliance because targets will not believe them (cf. Tedeschi, Schlenker, & Bonoma, 1973). Research suggests that people usually are concerned with establishing reputations for having high credibility, although they consider the opportunity costs of using punishments prior to issuing threats (Tedeschi, Horai, Lindskold, & Faley, 1970). Thus, a concern for managing impressions is often a factor in using punishments against other people. Felson (this volume) has presented evidence that physical assaults and homicides may often be triggered by verbal assaults by the victim on the actor's self-esteem. Violent reactions to verbal psychological provocations may have the function of punishing the violator of important social norms about propriety and mutual regard, particularly in close relationships.

Of course punishment may be wielded for a variety of other purposes. Physical force may be used to ward off an attacker or to take revenge. Self-defense and reciprocity norms may justify the use of noxious stimulation under some conditions. Punishment may be employed to somehow right the scales of justice by inflicting harm upon the perpetrator, as he did upon his victim. Sometimes justice norms are generalized by the harm-doer, who may interpret events in odd and idiosyncratic ways. For example, a woman in the United States during 1981 used her automobile to run over and kill a number of pedestrians in Las Vegas, Nevada. She said gamblers had kidnapped her daughter and alienated her affections, that her daughter was the bane of her life, and that she was taking revenge. In her mind, anyone on the streets of Las Vegas was a gambler, and thus, by faulty association, was responsible for the emotional loss of her daughter.

Punishment is often meted out to a particular person in order to set an example and deter others from acting in particular ways. There is even reason to believe that when punishment is certain and swift and legitimized, it is an effective deterrent (cf. Gibbs, 1968). The person or group may also justify punishment as a rehabilitative process. This idea may have originated with Christianity and the belief that suffering removes sin. However, there is precious little evidence that punishment builds great character.

An often neglected aspect of punishment is its symbolic function. Punishment administered through a judicial system, for example, conveys the values of the citizens of a community. The self-immolation of Buddhist monks in Indochina in the 1960s was symbolic of the suffering of the Vietnamese people and communicated protest against the war. Punishment meted out by supervisors or by police may represent attempts to restore authority and legitimacy in a social context (Toch, 1909; Kipnis & Misner, 1972). A person may use punishment as a way of restoring a sense of self-worth, especially when it is directed against a former oppressor (Fanon, 1963). Someone who is essentially alone, moving from one job to another or unemployed for long periods of time, unloved and without strong social ties to friends or family may need to do something to show himself or others that he exists, cannot be ignored, and can have an important impact on events. Lone political assassins often have these characteristics (Kirkham, Levy, & Crotty, 1970).

The use of punishments, particularly those involving strong noxious stimulation, tend to attract a great deal of attention. Television, cinema, and the news media provide vivid details of violent incidents, and people show great curiosity at the scenes of accidents. Terrorism also attempts to capitalize on the attention-getting feature of threats and punishments. The use of punishments for purposes of terrorism is a form of theater. The terrorist wants access to an audience in order to publicize his principles, cause, movement, or group. Without an audience there would be no reason for terrorism. An example of counterproductive reaction to terrorism was the reaction of the United States government and the print and electronic press to the kidnapping of Americans at the embassy in Iran. The terrorists could be assured of worldwide attention only as long as they did not set their captives free.

There is a saying about small children: You give them an inch and they will take a mile. This statement implies that a parent or guardian must be firm in maintaining control and discipline with a child. However, if we invert the example, we may note that sometimes an actor will use small amounts of force to explore the limits of his power. It is a common belief today that if the allied countries had been firm when Hitler's Germany invaded surrounding German speaking areas in the Late 1930s, world War II could have been averted. Hitler was encouraged by the lack of opposition by other countries to his expansionist policies and grew more and more bold in his actions. This kind of salami-slicing strategy can also be followed in interpersonal relations.

When identifying punishments, no assumption is made as to whether they are intentionally or deliberately administered. It is recognized that the production of harmful effects may be accidental, inadvertent, a mistake, or excusable for some other reason. These questions of causal attribution and legal or moral judgment arise after the identification of the relevant effects (Jones & Davis, 1965). It is another problem to explain why people have accidents, or why they make mistakes, and so on. Psychoanalytic theory alerts us to the possibility that people may be motivated to have accidents and to make mistakes.

Punishment may also be used as a tool of moral teaching. If we are to learn from the past, we cannot discount the large-scale impact of nonviolent tactics of disobedience such as used by Mahatma Gandhi and Martin Luther King, Jr. By accepting the punishment and even death as a consequence of protesting what are considered unjust laws, it is sometimes possible to engender guilt and withdrawal by those who possess superior material power. This is a complicated matter and little understood by social scientists. The work by Moscovici and Faucheux (1972) on how minoritics influence majorities is an importent first step in studying this kind of problem.

Further Comments on Threats and Punishments

A threat, whether contingent or noncontigent in form, may specify a particular form of punishment. Of course, threats may be transmitted without subsequent administration of punishment, and punishment may be meted out without prior issuance of a threat. The system of threats and punishments yields a taxonomy rich in descriptive possibilities. The terminology should be used with precision. Loose application of threat to mean threat to self-esteem or a target who feels threatened is not a correct application of a communication-oriented view of threats. A verbal statement containing negative evaluations of another person (e. g., name-calling) is a social punishment, not a threat, although the target may feel less secure because of it.

How does this social approach handle the problem of accidents? First, the form of punishment would be described. Then we would examine the criteria for holding a person legally or morally accountable for an event of that type. More relevant is how other naive people (not psychologists) assign responsibility and how they react to someone they hold responsible. Excuses and justifications may affect the attribution process.

It is this cycle of making moral judgments, labeling the behavior and actor (perhaps as aggressive), and then reacting to him that characterizes the research of Mummendey (this volume) and the group she has influenced in Western Europe. At the beginning of the cycle, however, is some form of threat or punishment, although it should be recognized that not all threats and punishments lead to this cycle. That is, the class of actions described as threats and punishments is larger than the subclass of actions labeled as aggressive. Whenever the use of threats and punishments is perceived as legitimate or justifiable or excusable, the action and the actor will not be perceived as aggressive.

Does the language of threats and punishments allow social scientists to identify the actions that theorists of aggression are interested in studying? It would be difficult to think of an action that might be labeled as aggressive that would not fit into the taxonomy. We have avoided the inclusion of all behaviors, as would Bandura's definition of aggression as a skill, but have also avoided the problems attendant upon the behavioristic and attributional definitions.

Conclusion

In summary, observation, evidence, and intuition suggest that threats and punishments may be used by persons to gain compliance from targets; for impression management purposes; as a means of gaining attention from an audience; for reasons of self-esteem; for self-defense; for reciprocity; to establish justice; as a form of rehabilitation; as a commitment strategy; as a symbolic expression of adherence to values or systems of authority; and as a tactic exploring the limits of power and influence.

A compilation of these explanations for the use of threats and punishments does not, of course, constitute a theory but only a prolegomenon to one. What the social scientist would like to know is when, under what circumstances, by whom, against whom, and in what form threats and punishments will be used. Allied questions include how observers perceive the wielder of coercive power and when it will be labeled as aggression, when and in what way an actor may attempt to absolve himself of responsibility for harmful effects of behavior (e. g., excuses), and what kinds of legitimations and justifications are given.

Answering the above questions is surely a very complex and important undertaking, but the job will not get done if social scientists refuse to disencumber themselves of the archaic and value-laden language serving as the basis for most current psychological theories of aggression. Struggling to make distinctions between hostile and instrumental aggression and observing the conditions under which various physical and technical skills are learned do not enlighten us about the harm people do to one another, because they are responses to the wrong questions. Words like threats, punishments, and coercion are clearly interaction terms that would have no meaning outside of a social context. Hence, the observational terms proposed herein have the value of focusing the theorists on the social determinants of the behaviors to be studied.

Tedeschi (1983), has proposed that perspectives emphasizing inner biological and psychological processes, such as instincts, hormones, brain centers, frustration, and unconscious motives, are inadequate and often irrelevant for understanding why people use threats and punishments. While the evidence is not compelling, it may be true that on occasion aberrant and violent behavior may be attributed to some inner state of the person, but for the far greater number of incidents involving threats and punishments, the determinants will be social and interactional in nature.

Laboratory Research on Aggression

If observational constructs of aggression are inadequate in designating any specific set of responses as aggression, then what can be said about laboratory research on aggression? What counts as aggression in such studies? Among the dependent variables in this area of research are: punching an inanimate object constructed especially for pummeling; word associations considered by the experimenter as hostile; slightly unfavorable ratings of others; delivering noises, shocks, and bright lights to target persons; and reports of fantasies and dreams. All of these responses are made in the context of rules and norms established for the situation by the experimenter, mostly for middle class American college students.

Norms, Justification, and Harm-doing

We must consider the effects of legitimating the behavior of subjects, particularly when generalizing results to other contexts. It may be the case that laboratory research on aggression lacks both external and ecological validity. Subjects are given cover stories and legitimations for performing the "aggressive" behavior, but of course when they are specifically told the experimenter is interested in aggression, they are inhibited from performing the relevant behavior (Diener, Dineen, Endresen, Beaman, & Fraser, 1975). That is, subjects in Laboratories only perform behaviors that have been legitimated for them and where they believe no harm will actually occur to the victim, or whatever harm is to be delivered will be outweighed by the benefits to be achieved.

One way to view this area of research is that it focuses on the use of noxious stimulation by subjects who have had its use legitimized for them by the experimenter. But what is it that aggression researchers want to explain? In the beginning, the observation was that people harm or try to harm one another frequently enough to make it an important social problem. On many occasions a consensus exists among observers that the use of threats and/or punishment is not justified (or excusable). The illegitimate and antinormative use of threats and/or punishment is perceived as aggression if the observer attributes intent to the actor (Brown & Tedeschi, 1976). Since the behavior of Laboratory subjects is legitimated, it would be expected that naive observers would not view their behavior (e. g., use of shocks, noise, negative evaluations, etc.) as aggressive. Indeed, subjects are not perceived as aggressive in the Berkowitz (Berkowitz, Green, & Macaulay, 1962) essay-writing paradigm (Kane, Joseph, & Tedeschi, 1976). Bandura's (1973) modeling paradigm (Joseph, Kane, Nacci, & Tedeschi, 1977), or Taylor's (1967) competitive reaction-time game (Stapleton, Joseph, & Tedeschi, 1978). Thus, without an adequate observational definition to allow for a clear identification of the relevant behaviors and with no consensus among observers (lay and scientific) about how to label such responses, there may be some question about what is being studied.

In many instances it seems clear that some form of punishment is the focus of aggression research. Sometimes the punishment takes the form of deprivation of resources, at other times social punishment, but most often it is a deception that leads subjects to believe they are using noxious stimulation. In most studies the subject is attacked, insulted, or otherwise abused or mistreated by another person, usually a confederate in the employ of the experimenter. Social norms of self-defense, reciprocity, and social justice become relevant under these conditions and justify some form of harm-doing by the subjects, particularly since the experimenter provides additional justification by saying the harm-doing is to be interpreted some other way, such as evaluation of an essay, the administration of punishment to aid the learning process, or simply the production of stress so that scientific study of it can be carried out. As Mixon (1972) has shown, if subjects are told they are to harm another person without assurances or legitimations, 100% of them refuse to do so.

External and Ecological Validity of Laboratory Research

Aggression theorists set out to understand the factors that cause people to assault, murder, rape, rob, and otherwise harm one another. Most of the actions of interest are considered antinormative or illegitimate. The question raised by these observations is: Are the factors that increase or decrease legitimated harm-doing in the laboratory the same as antinormative harm-doing outside of it? This is of course an empirical question, but given the complex factors considered earlier for the use of threats and punishments, one must remain skeptical about the ecological validity of laboratory experiments or the applicability of the theories based on them. The research focus seems to be on a small segment of the universe of relevant actions.

Social psychologists have managed to study antinormative behavior, such as cheating, lying, and stealing (Hartshorne & May, 1928). However, normative considerations have been ignored by aggression researchers. Their theories do not sensitize them to such social factors. If the theories are to be seriously used to explain antinormative conduct, it will be necessary for studies to be conducted in situations where the relevant dependent variable is clearly an illegitimate behavior. The concern would be for the external validity of the results obtained under legitimated conditions. Subjects, of course, are reluctant to harm others without legitimation and that is why psychologists provide it. We will need to be clever if we are to induce illegitimate conduct under controlled conditions. Perhaps we will be limited to nonlaboratory studies.

It should be noted that there are also laboratory studies of threats (cf. Tedeschi et al., 1973). Seldom is the word aggression used in the threat literature. There is clearly much to learn about threats. For example, if a model typically uses threats of social punishments, would an observer imitate by specifically using the same form of threat or would the observer tend to more frequently use all kinds of threats? Laboratory research should be done with more precision than the old language of frustration and aggression allows. Consideration of the relevant phenomena as various forms of threats and punishments provides a more articulate and precise observational vocabulary for research.

Conclusion

Linguistic analysis of observational constructs of aggression reveals that no available definition allows one to discriminate between aggressive and nonaggressive behaviors. Furthermore, the need to make attributions of intention and assessments of excuses and justifications reduces the so-called behavior to a problem of moral judgments. Concepts of threat and punishment have been offered as a basis for a new approach to developing theory and research. The new terminology not only provides a substitute language for the old discarded one, but also refers to a much larger universe of social behavior than has been typically the domain of aggression theorists. Aggression researchers have typically restricted themselves to nonaccidental behaviors used in a legitimated laboratory context where harm-doing is instigated by antecedent actions of a confederate.

A theory of coercive power (e.g., Tedeschi, Schlenker, & Lindskold, 1972) should consider all occurrences of the relevant events, whatever the reason for them. If the past research on aggression is to be used as a basis for explaining antinormative actions, new research will need to establish the validity of doing so. This new research may of necessity have to be performed outside the laboratory, given the difficulty of eliciting antinormative conduct from subjects who know they are under surveillance.

It seems likely that in the future we will need to stress norms, systems of justification, and excuses and the role of emotions, attributions of intent, and the labeling of actions and actors, as well as social power, when explaining the use of threats and punishments. This social interactionist approach will lead us away from the intrapsychic and biological perspectives on aggression toward a set of new problems, theory, and research.

References

Anscomb, G. E. M. *Intention*. Oxford, England: Blackwell, 1963.

Bandura, A. *Aggression: A social learning analysis*. Englewood Cliffs, N.J.: Prentice-Hall, 1973.

Berkowitz, L., Green, J. A., & Macaulay, J. R. Hostility of the scapegoat. *Journal of Abnormal and Social Psychology*, 1962, *64*, 293–301.

Birdwhistle, R. I. Some relationships between American kinesics and spoken American English. In A. G. Smith (Ed.), *Communication and culture*. New York: Holt, Rinehart & Winston, 1966.

Brinton, C. *The anatomy of revolution* (rev. ed.). Englewood Cliffs, N.J.: Prentice-Hall, 1952.

Brown, R. C., Jr., & Tedeschi, J. T. Determinants of perceived aggression. *Journal of Social Psychology*, 1976, *100*, 77–87.

Buss, A. H. *The psychology of aggression*. New York: Wiley, 1961.

Diener, E., Dineen, J. Endresen, K., Beaman, A. L., & Fraser, S. C. Effects of altered responsibility, cognitive set, and modeling on physical aggression and deindividuation. *Journal of Personality and Social Psychology*, 1975, *31*, 328–337.

Dollard, J., Doob, L. W., Miller, N. E., Mowrer, O. H., & Sears, D. R. *Frustration and aggression*. New Haven, Conn.: Yale University Press, 1939.

Fanon, F. *The wretched of the earth*. New York: Grove, 1963.

Feshbach, S. The drive-reducing function of fantasy behavior. *Journal of Abnormal and Social Psychology*, 1955, *50*, 3–11.

Freud, S. *A general introduction to psycho-analysis*. New York: Boni & Liveright, 1920.

Gibbs, J. P. Crime, punishment, and deterrence. *Southwestern Social Science Quarterly,* 1968, *48,* 515–530.

Hartshorne, H., & May, M. A. *Studies in the nature of character.* Vol. 1. *Studies in deceit.* New York: Macmillan, 1928.

Heider. F. *The psychology of interpersonal relations.* New York: Wiley, 1958.

Holsti, O. R. Content analysis. In G. Lindzey, & E. Aronson, *The handbook of social psychology* (Vol. 2). Reading, Mass.: Addison-Wesley, 1968.

Jones, E. E., & Davis, K. E. From acts to dispositions: The attribution process in person perception. In L. Berkowitz (Ed.), *Advances in experimental social psychology* (Vol. 2). New York: Academic Press, 1965.

Jones, E. E., & Pittman, T. S. Toward a general theory of strategic self-presentation. In J. Suls (Ed.), *Psychological perspectives on the self.* Hillsdale, N. J.: Erlbaum, 1982.

Joseph, J. M., Kane, T. R., Nacci, P. L., & Tedeschi, J. T. Perceived aggression: A re-evaluation of the Bandura modeling paradigm. *Journal of Social Psychology,* 1977, *103,* 277–289.

Kane, T. R., Joseph, J. M., & Tedeschi, J. T. Person perception and an evaluation of the Berkowitz paradigm for the study of aggresssion. *Journal of Personality and Social Psychology,* 1976, *33,* 663–673.

Kipnis, D., & Misner, R. P. *Police actions and disorderly conduct.* Mimeographed manuscript, Temple University, 1972.

Kirkham, J. S., Levy, S., & Crotty, W. J. (Eds.), *Assassination and political violence.* New York: Praeger, 1970.

Mixon, D. Instead of deception. *Journal for the Theory of Social Behaviour,* 1972, *2,* 145–177.

Moscovici, S., & Faucheux, C. Social influence, conformity bias, and the study of active minorities. In L. Berkowitz (Ed.), *Advances in experimental social psychology* (Vol. 6). New York: Academic Press, 1972.

Ryle, G. *The concept of mind.* London: Hutchinson, 1949.

Stapleton, R. E., Joseph, J. M., & Tedeschi, J. T. Perceived aggression and competitive behavior. *Journal of Social Psychology,* 1978, *105,* 277–289.

Taylor, S. P. Aggressive behavior and physiological arousal as a function of provocation and the tendency to inhibit aggression. *Journal of Personality,* 1967, *35,* 297–310.

Tedeschi, J. T. Threats and promises. In P. Swingle (Ed.), *The structure of conflicts.* New York: Academic Press, 1970.

Tedeschi, J. T. Social influence theory and aggression. In R. Geen, & E. Donnerstein (Eds.), *Aggression: Theoretical and empirical reviews.* New York: Academic Press, 1983.

Tedeschi, J. T., Gaes, G. G., & Rivera, A. N. Aggression and the use of coercive power. *Journal of Social Issues,* 1977, *33,* 101–125.

Tedeschi, J. T., Horai, J., Lindskold, S., & Faley, T. E. The effects of opportunity costs and target compliance on the behavior of the threatening source. *Journal of Experimental Social Psychology,* 1970, *6,* 205–213.

Tedeschi, J. T., & Melburg, V. Aggression as the illegitimate use of coercive power. In Blumberg, Hare, Kent, & Davies (Eds.), *Small groups* (rev. ed.). London: Wiley, 1983.

Tedeschi, J. T., Melburg, V., & Rosenfeld, P. Is the concept of aggression useful? In P. Brain & D. Benton (Eds.) *A multi*-disciplinary approach to aggression research. Elsevier North Holland, Biomedical Press, 1981.

Tedeschi, J. T., & Riess, M. Verbal tactics of impression management. In C. Antaki (Ed.) *Ordinary language explanations of social behaviour.* London: Academic Press, 1981.

Tedeschi, J. T., Schlenker, B. R., & Bonoma, I. V. *Conflict, power, and games.* Chicago: Aldine, 1973.

Tedeschi, J. T., Schlenker, B. R., & Lindskold, S. The source of influence: The exercise of power. In J. Tedeschi (Ed.), *The social influence processes.* Chicago: Aldine, 1972.

Tedeschi, J. T., Smith, R. B., Ill, & Brown, R. C., A reinterpretation of research on aggression. *Psychological Bulletin,* 1974, *89,* 540–563.

Thorndike, E. L. *The elements of psychology.* New York: A. G. Seiler, 1905.

Toch, H. H. *Violent men: An inquiry into the psychology of violence.* Chicago: Aldine. 1969.

Chapter 2

Motivation Theory of Aggression and Its Relation to Social Psychological Approaches

Hans-Joachim Kornadt

While studying the recent situation of aggression theory and its development, we can observe a shift of approaches. At first, individual factors (like drive, instinct, or trait) were considered, then general processes (e. g., the frustration-aggression relationship or imitation), and, finally, situational factors and thereby more and more their subjective evaluation were stressed.

But the majority of researchers seems to be dissatisfied with the situation of theoretical understanding. For example, Albert Pepitone wrote:

[Firstly.] . . . the behavioristic emphasis on the *overt* responses resulted in a neglect of the internal states underlying aggressive acts such as anger, rage and hostile feelings. [Secondly,] . . . the conception of aggression as a *reaction* to frustration led theorists to ignore "impulsive" aggression . . . [Thirdly,] the definition of aggression as a goal response severely restricted the domain of research and theory. . . . Some aggression committed in uncontrollable circumstances . . . should be addressed by the theory, even though such "expressive" aggression is not a directed goal response. . . . Further the requirement that aggression be a goal response also disqualified from theoretical consideration "instrumental" actions, that are not driven by an aggressive instigation but that nonetheless may have injurious consequences for those to whom they are directed. . . . [And finally,] . . . in avoiding mentalistic assumptions, the frustration-aggression hypothesis failed to take account of cognitive factors in aggression; the objective concepts in the theory cannot deal with effects that are based on the *meaning* of the frustration. (Pepitone, 1981, p. 3 f.) Recently at the Cross-Cultural Conference in Aberdeen, Pepitone added biological factors as another field also neglected by psychological theories (Pepitone, 1982).

On the whole, I agree with Pepitone's criticism. Surely we have to broaden the view in several directions, though perhaps not in all of those Pepitone mentioned. We have to take into account:

1. *Biological factors,* at least in the development of human aggressiveness. We cannot ignore some biological roots.
2. *Internal processes,* especially affective and cognitive ones.

3. *Reactive* as well as *spontaneous* forms of aggression.
4. Not just *goal-directed* forms of aggression, but also others.

In contrast to Pepitone, I would include *impulsive aggression,* but not *instrumental aggression.*

Regarding the last point, I think we should not broaden our view too much at once. We should rather focus at first on clear, aggression-specific aspects. If we try to explain all forms of behavior which somebody might call "aggression," then we would include a far too heterogeneous variety of so-called aggression. Most likely all of these behaviors do not follow the same rules. Therefore, we cannot hope to settle on one single aggression theory which is able to explain all these phenomena at once. But we must use criteria from which a *homogeneous* class of aggression can be constituted.

On the other hand, another – often neglected – aspect exists, and should be included: namely the *individual differences* in aggressiveness. We know that these differences are effective, that is quantitative and qualitative. And – perhaps even more important – we know also that they are amazingly *stable.* Both statements are empirically supported, e.g., by Olweus (1979, 1980; see also Loeber, 1982). That leads to the idea that perhaps the systematic consideration of those differences can bring us toward a new theoretical view. The stability of individual differences in aggressiveness is an indicator of an *enduring disposition* to behave aggressively. Such a disposition cannot, of course, simply be an aggressive trait or "instinct," which operates regardless of situational factors. It can only be a complex, more or less individually organized system which interacts with situational factors. Understanding more of its organization may enable us to understand better the peculiarity of aggressive behavior and to differentiate aggression from other kinds of behavior. In pursuing Pepitone's criticism and summarizing the central points under consideration, I want to state some postulates:

1. Aggression research should start by recognizing the fact that sometimes people act out of the intention to harm others or to be destructive, and are satisfied only by having done so; the intention arises not only in reaction to perceived violation of norms. These facts point to a class of phenomena which need particular explanation.
2. Aggression theory should recognize as a basic fact that there is in all higher animals a universal behavioral system to attack others of the same species under defined conditions, especially aversive conditions (attack, threat, or rivalry, and if flight is impossible or not yet necessary). Under those conditions rage overrides fear, and in a way rage-based and anxiety-based systems function exclusively.
3. The basic characteristics of such a system should likewise be effective in the beginning of the development of human behavioral systems. (The inference from animal to human behavior does not go further than to establish a *hypothesis* concerning the starting conditions of human developmental processes.)
4. Taking anger as a prominent and central part from the beginning and – at a much later stage – hostile intentions as typical elements of aggression, then this behavioral system need not be arbitrarily or speculatively defined and discriminated from others, since it consists of a functional system.

5. It is then a challenging and promising task for research to study:
   The process of development
   The structure of aggressive behavior at different developmental stages
   The final structure of the behavioral system (the disposition for aggression)
   The specific relationship between intention and moral judgment
6. More attention should be paid to individual differences. They are an important source of variance in aggressive behavior and, moreover, are an important
   indicator of the function and structure of the disposition to aggression.
7. Aggression research in line with these postulates should (and can) be done
   within the framework of a *specific* theory.

A theoretical approach which seems to be suitable for comprehending and integrating all the aspects mentioned is a motivation theory, as developed by
McClelland, Atkinson, Heckhausen, Fuchs, Weiner, and many others. If we
compare aggression with other kinds of motivated behavior, e. g., achievement-,
power-, sex-, or anxiety-motivated behavior, then it can hardly be doubted that
aggression does in fact have its own basic peculiarities and regularities despite
some overlappings with other types of behavior. Since the concept of "achievement motive" as a hypothetical construct was useful and stimulative in achievement research (that is, in differentiating achievement from other behavior), why
should it not likewise be successful for interpreting the disposition to aggression
as an enduring aggression motive? A motive is understood as an affective-based
behavioral system (McClelland, Atkinson, Clark, and Lowell, 1953; Fuchs, 1963;
Heckhausen, 1968, 1972, 1980; Kagan, 1972; Kornadt, Eckensberger, and Emminghaus, 1980). With the evidence that anger is most likely the specific biological base for the development of aggressiveness, it seems indeed possible that a
motive should constitute the functional unit we are looking for. That idea was
the central hypothesis we started from many years ago. In the meantime the idea
was elaborated into an empirically founded theory (Kornadt, 1974, 1982). I shall
attempt to describe it briefly, as it is impossible to present all of the detailed arguments and findings.

Motivation Theory of Aggression

Basic Assumptions

The basic assumptions were derived from the general motivation theory as a
frame of reference, as it was developed by McClelland et al. (1953), Atkinson
(1958, 1964), Atkinson and Feather (1966), Weiner (1972), Heckhausen (1968,
1972, 1980), and Fuchs (1963, 1976). It has a broad empirical basis mostly in
achievement behavior, and it offered basic features of an universal motivation
theory (Kornadt et al., 1980), as well.

In achievement behavior, according to the formula of Atkinson (1964), the
motivation for a specific achievement act (motivat$_A$) is seen as a function of two
motivational components: an approach and an avoidance motive. The approach
motive consists of the enduring achievement motive (M_A), the expectancy of

achieving success by the specific act, and the incentive of the anticipated goal (E_E and I_A); the same factors exist for the avoidance component (fear of failure). Starting with an analogous hypothesis for aggression, we assumed an aggression motive as the approach component and inhibition as the avoidance component. Their functioning can be seen accordingly. The motivation for a specific aggressive act is a function of the enduring aggression motive ($M_{aggr.}$), the expectancy to be successful, and the incentive of the aggression ($E_{aggr.}$); ($I_{aggr.}$), minus the enduring motive to avoid aggression ($M_{aAggr.}$), the expectancy to be punished and the negative incentive of being punished (E_p; I_p). This can be expressed in the hypothetical formula

$$\text{Motivation}_{spec.aggr.} = f(\underbrace{M_{Aggr.} \times E_{Aggr.} \times I_{Aggr.}}_{\text{approach}}) - (\underbrace{M_{aAggr.} \times E_p \times I_p}_{\text{avoidance}})$$

The central ideas are the assumption of 1. the two enduring motive components ("aggression motive" and "aggression inhibition"), and 2. their interaction with the specific conditions of the given situation in terms of expectancy and value (incentive).

The particularities of a motive are (with reference to Heckhausen, 1963, 1968) defined by the specifity of the cognitive and emotional properties of events which are intended. That is, in the case of aggression, the intention of destructive, injurious effects on others in order to remove sources of frustration. While so far this is not more than a very rough description which needs much more elaboration, even this approach has an important consequence, namely, in respect to the old problem of definition. The motive theory includes the hypothesis that there is a section of the heterogeneous universe of "aggressive" behavior for which a specific aggression motive is the central functional factor. Consequently, all these specifically motivated, hostile aggressions are basically organized by the functional regularities of this motive. One obtains, therefore, a functional criterion for the differentiation of specific, (intended, hostile) aggressive acts and those acts where injuries as such are not intended, but may arise as side effects in the pursuit of nondestructive goals.

A Detailed Analysis of Sequence of Action

To become fruitful for further research an elaboration of the theory is necessary in three aspects:

In aggressive behavior, how does the motive interact with the specific situation and which different elements of the enduring components are involved?

In what functional parts of the motive can individual differences be seen?

By what processes is such a motive developed in general, and what processes and influences would produce individual differences?

This paper can only deal with the first point. To get more detailed hypotheses, one has to analyze the process of an actual aggressive act. Within the theoretical framework, it must be explained how the latent enduring motive is activated, what are the internal consequences leading to a specific aggressive act, and what are the conditions for a final deactivation of the motivation, the ending of the act. The whole process and its different steps are illustrated in Fig. 2.1.

The diagram illustrates the hypothetical sequence of a "normal" case according to general assumptions concerning motivated acts.

At first, there must be some situational cues leading to an activation of the motive. It is assumed that normally these are "frustrating" conditions which activate primarily anger. The connection between "frustrating" cues and anger is seen as twofold: as an inborn capacity to react with specific affective arousal (anger) to particular cues or events (e.g., Berkowitz, 1969) and as a learned connection. It is important that neither "frustration" nor anger is seen as leading immediately to aggression – at least not in regular cases (perhaps in "impulsive" aggression). Anger and arousing cues of the situation will be subjects of cognitive interpretation and control, and only if a situation is subjectively seen as "really annoying" will the enduring aggression motive become activated. This activation will not occur if the situation is "funny" or "neutral."

Activation means the actualization of generalized goals, of patterns of instrumental behavior, and connected "expectancy emotions" (Heckhausen, 1980). Mainly, one may understand the enduring motive as consisting of these in particular. Then these highly generalized cognitive schemata, patterns, and emotions need to be specified and applied to the given situation. Here, especially the expectance x value theory comes in to play. Situation-related goals and appropriate acts must be developed and the incentive of goal attainment and the probability of being successful by the acts in question have to be considered. At the same time the negative component of the aggression motive system, the aggression inhibition, can be activated. In that case, actualized anticipations of possible negative consequences of aggression (guilt, punishment) are also taken into account.

The next step is a decision – whether or not to set a specific aggressive goal. In case of deciding *for* an aggressive goal, this goal will normally have a very situation-specific cognitive structure. For example, if I am angry at colleague A, since he has offended me in public, then I will be satisfied not simply by aggression against anyone, but only by taking "revenge" against A; even this needs an appropriate form and situation, e.g., similarly in public. Then the instrumental act needed to attain the goal will be decided on, and it will be carried out under appropriate conditions. That may comprise the decision to delay the act until a suitable occasion appears.

The outcome of the act will again be interpreted in terms of whether the intended effect is reached or not. In the first case, the consequence will be a feeling of relief, of being satisfied, proud, or the like, and this will result in a deactivation of the previously activated motive. Otherwise the motivation remains activated and the intention is still relevant. This is of importance for the question of catharsis as well as "spontaneous" aggression (Kornadt, 1974, 1982). Thus, the steps in the aggressive sequence may be described.

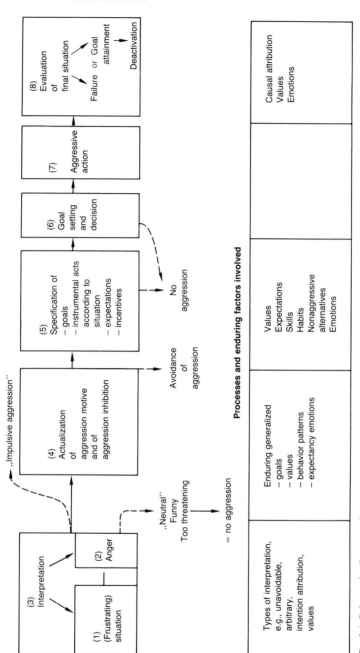

Fig. 2.1. Schematic diagram of the sequence of an aggressive act

Relationship Between Motivation Theory and Social Psychological Approaches

Aggressive behavior is normally socially interactive, interpersonal behavior. As mentioned at the outset, it has therefore been studied mostly from a social psychological point of view. That means that primarily overt behavior and its external antecedents will be the object of an analysis "from the outside." The behavior may be seen as "aggressive" by a victim or a neutral observer, neither of whom knows the subjective meaning and motivation of the "aggressor." But a problem remains: the term "aggression" as a label for the behavior studied refers to heterogeneous kinds of behavior and loses its precision and special meaning.

Motivation theory pays more attention to the internal conditions and processes in trying to explain the internal organization of behavior and its motivational reasons. Therefore, it studies primarily the individual side of the interaction and can, therefore, not avoid becoming more aware of individual differences. But this does not mean that the two approaches are mutually exclusive or contradictory. On the contrary, they complement each other.

At the "microlevel," social psychologists may study, for example, the relationship between certain interpretations of another person's behavior and the aggressive reaction in general or the relationship between the perceiving of another's behavior as ruthless or aggressive and the aggressive reaction in general, perhaps as the justification for one's own aggression ("aggressive is always somebody else," Mummendey, Bornewasser, Löschper, and Linneweber, 1982). It studies the same phenomena as motivation theory in field 3 of the model of the sequence (Figure 2.1). Social psychologists may also study the situational conditions, social or cultural influences on the interpretation focusing on the general relationship.

Motivation theory, on the other hand, additionally studies the fact that within the person there also exist enduring tendencies and individual differences in interpreting certain events in a specific way.

At the "macrolevel," social psychologists are interested in "aggressive" interactions (Mummendey, 1982), for instance how an aggressive escalation occurs. They try to describe whole interactive sequences, including actions, their dissenting interpretations (that is, each participant sees the other as violating norms and feels subjectively justified), respective reactions, and so forth. Here again general rules may be discovered and described.

We believe that those theoretical explanations cannot cover the whole range of phenomena, since not all persons behave in the same way. Aggressive interactions can be only insufficiently described by general rules, though it is of course necessary to know them. Generalizations must neglect individual differences in the tendency to behave aggressively and – even more important – such general rules can never explain how those differences are developed.

Motivation theory, as previously described, is one-sided in another way. It concentrates primarily on one side of the interaction, taking into account one individual with his specific enduring personal tendencies, beliefs, affects, goals, and so on. Thus, the chain of interactive sequence can be explained from the view of one partner, including why A becomes involved in an aggressive inter-action with B, and why – even under the same conditions – this is not the case for C.

In addition, some social psychologists and motivation theorists are studying what are probably different phenomena. An aggressive interaction essentially determined by situational conditions and based on dissenting interpretations, mutual insults, and reproaches and self-justifications is not the same type of thing as a rage-based or even spontaneous aggression. While the first can be an interaction between equally involved persons with exchangeable positions (see Mummendey, 1982, p. 330), the latter two are interactions with the aggressor much more active in structuring the situation. Moreover, the first case belongs more to the instrumental type of aggression, while the latter two belong to the hostile, specifically motivated type of aggression.

Only the combination of both approaches and their application to various in-teractions between various persons would be able to describe and explain whole interactive sequences. One should be aware that this is a problem of "higher order complexity."

Even if one agrees that the different elements of the aggression motive men-tioned are relevant and functioning, one may doubt whether they really combine to form such a motivational system. And one may ask – like Tedeschi – what makes the "glue" that combines the different elements into a "system." From a motivation theory point of view, the answer is simple: Here we have to refer to the development of the motive system. The glue comes from the individual de-velopmental process. During that process specific connections are formed among the different elements. Certain situational conditions will be connected, e. g., with the tendency to prefer the assumption of unfriedly intentions and thus provoke anger, while other cues are connected with neutral interpretations; or anger will become individually connected with specific patterns of behavior and certain expectancy emotions, and these in turn are connected with specific val-ues, and so forth. Therefore, the system is individually organized and results from countless individual experiences. The process of development and the fac-tors which are influencing it in general and which may lead to individual differ-ences in aggressiveness are described elsewhere according to the motivation theory (Kornadt, 1981, 1982).

Essentials of the Motivation Theory

After such an *excursus* the essentials of the motivation theory of aggression shall be summarized. The essentials of the motivation theory of aggression are:
   1. Aggressive behavior results from an interaction of situational and enduring personal factors.
   2. The enduring personal factor is understood as an aggression motive system. It consists of two components: aggression motive and aggression inhibition.
   3. The enduring aggression motive consists of a number of different elements, e. g.

An enduring connection between frustrative cues and anger:
- Enduring styles of interpretation
- Enduring generalized goals and behavioral patterns
- Enduring values, effective in the interpretation of frustration as well as in the incentive of goals considered
- Enduring expectancy emotions
  This is also the case with the aggression inhibition.
4. A systematic relationship between the different and relatively stable elements constitutes the motive.
5. In those enduring elements individual differences will be found.
6. Specific processes bring about and influence the development of the motive system and of individual differences.
7. As a consequence of these points, the motive construct can be tested in various independent ways:
- In its direct and global behavioral effects
- With respect to the assumed individual differences
- With respect to the detailed elements assumed and the systematic relationship among the various factors
- And most important, with respect to the developmental conditions. Therefore the danger of circularity of the definition, as in the frustration-aggression theory, is avoided.
8. The detailed hypotheses about the development of the motive include the assumption of a biologically-based relationship between aversive cues and anger, further hormonally caused sensitization of the brain, learning dispositions, etc. These are considered as the universal starting point of developmental processes and as the first simple functional core of the motive. But these *do not directly influence* the final aggressiveness. They are only the prerequisite and/or the basis for many learning processes, taking place during experiences of child-rearing, in the playground, in school, etc. The motivation theory implies specific hypotheses, which developmental conditions lead to the development of specific parts of the aggression motive system in general, and from specific experiences specific individual differences, even culture-specific qualities of the motive are predicted (see Kornadt et al., 1980, Kornadt, 1983).
9. Referring to the second point of Pepitone's criticism (that is, aggression is only seen as reactive): Motivation theory refers to and also explains *spontaneous forms* of aggression. The decision to strive toward attaining a certain aggressive goal may include some delay, like waiting for the best occasion (Lambert, 1974, 1982) or a highly generalized goal like, "I hate all policemen and I will hurt one whenever possible" includes the intention to look for appropriate occasions.
10. The theory predicts that the negative affect of anger is not the only affective base. *Pleasurable affects* can also be connected by learning with cues evoking positive anticipations of committing aggressive acts and of reaching aggressive goals (e. g., a pupil, observing a mistake by the teacher, may pleasurably plan and then execute acts to annoy the teacher).

11. Finally, not always the described form of intended aggressive act must be assumed. Under certain circumstances, *"impulsive"* aggression or "push-type behavior" can occur according to the theory. For example, under strong affects the complicated cognitive processes of evaluation and control may be simplified, shortened, or even dropped.

Conclusion

The motivational approach is sometimes misunderstood as a simple individualistic approach comparable with the old trait or instinct theories. I very much hope that I can make clear that this is not the case. From the beginning of McClelland's or even Murray's work (1938), the concepts of motive and of motivation were always understood as functioning only in interaction with situational conditions, as they are subjectively perceived.

Every motive is a system of enduring dispositions which in themselves are developed in interaction between person and environment, with needs, hopes, and fears on the one hand, and incentives, threats, and positive and negative rewards on the other hand.

The activated motive, the motivation, is always and only adapted to the actual situation as a result of many interactive processes, such as selectively processing information or evaluating the relevant situational circumstances and rejecting all possible goals and means which are not suitable in the situation.

Therefore, I am convinced that neither the pure individualistic approach, as in the former trait or instinct theories, nor the pure social psychological approach, since it must focus on situational factors and general rules as if all persons would behave in the same way, can be correct. A motivation theory of aggression can avoid the shortcomings of both of these extreme viewpoints, and is the only opportunity to understand which peculiarities aggressive behavior constitute.

References

Atkinson, J. W. *Motives in fantasy, action, and society.* Princeton, N. J.: VanNostrand, 1958.

Atkinson, J. W. *An introduction to motivation.* Princeton, N. J.: VanNostrand, 1964.

Atkinson, J. W., and Feather, N. T. *A theory of achievement motivation.* New York: Wiley, 1966.

Berkowitz, L. *Roots of aggression.* New York: Atherton Press, 1969.

Fuchs, R. Funktionsanalyse der Motivation. *Zeitschrift für experimentelle und angewandte Psychologie,* 1963, *10,* 626–645.

Fuchs, R. Furchtregulation und Furchthemmung des Zweckhandelns. In A. Thomas (Ed.), *Psychologie der Handlung und Bewegung.* Meisenheim am Glan: Anton Hain, 1976, 97–162.

Heckhausen, H. Eine Rahmentheorie der Motivation in zehn Thesen. *Zeitschrift für experimentelle und angewandte Psychologie,* 1963, *10,* 604–626.

Heckhausen, H. Achievement motive research: Current problems and some contributions towards a general theory of motivation. In W. J. Arnold (Ed.), *Nebraska symposium on motivation.* Lincoln, Neb.: University of Nebraska Press, 1968, 103–174.

Heckhausen, H. Die Interaktion der Sozialisationsvariablen in der Genese des Leistungsmotivs. In C.J. Graumann (Ed.), *Handbuch der Psychologie,* Band 7, Teil 2: *Sozialpsychologie.* Göttingen: Hogrefe, 1972, 955–1019.

Heckhausen, H. *Motivation und Handeln.* Berlin: Springer, 1980.

Kagan, J. Motives and development. *Journal of Personality and Social Psychology,* 1972, *22,* 51–66.

Kornadt, H.-J. Toward a motivation theory of aggression and aggression inhibition: Some considerations about an aggression motive and its application to TAT and catharsis. In J. de Wit, and W.W. Hartup (Eds.), *Determinants and origins of aggressive behavior.* The Hague: Mouton, 1974, 567–577.

Kornadt, H.-J. Development of aggressiveness: A motivation theory perspective. In R.M. Kaplan, V.J. Konečni, R.W. Nocaco (Eds.) Aggression in children and youth. The Hague: Nijhoff, 1984, 73–87.

Kornadt, H.-J. Aggressionsmotiv und Aggressionshemmung. *Empirische und theoretische Untersuchungen zu einer Motivationstheorie der Aggression und zur Konstruktvalidierung eines Aggressions-TAT* (Band 1). Bern: Huber, 1982.

Kornadt, H.-J. *Erziehung und Aggression im Kulturvergleich.* Bericht über die im Rahmen der Voruntersuchung und mit Mitteln der Stiftung Volkswagenwerk durchgeführten Arbeiten vom Februar 1980 bis Oktober 1982. Saarbrücken, 1983 (unpublished manuscript).

Kornadt, H.-J., Eckensberger, L.H., and Emminghaus, W.B. Cross-cultural research on motivation and its contribution to a general theory of motivation. In H.C. Triandis, and W. Lonner (Eds.), *Handbook of cross-cultural psychology. Basic processes* (Vol.3). Boston: Allyn and Bacon Inc., 1980, 223–321.

Lambert, W.W. Promise and problems of cross-cultural exploration of children's aggressive strategies. In J. de Wit, and W.W. Hartup (Eds.), *Determinants and origins of aggressive behavior.* The Hague: Mouton, 1974, 437–460.

Lambert, W.W. *Some strong strategies in the aggression of children in six cultures.* Cornwall University, 1982 (unpublished manuscript).

Loeber, R. The stability of anti-social and delinquent child behavior: A review. *Child Development,* 1982, *53,* 1431–1446.

McClelland, D.C., Atkinson, J.W., Clark, R.A., and Lowell, E.L. *The achievement motive.* New York: Appleton-Century-Crofts, 1953.

Mummendey, A. Zum Nutzen des Aggressionsbegriffs für die Aggressionsforschung. In R. Hilke, and W. Kempf (Eds.), *Aggression. Naturwissenschaftliche und kulturwissenschaftliche Perspektiven der Aggressionsforschung.* Bern: Huber, 1982, 317–333.

Mummendey, A., Bornewasser, M., Löschper, G., and Linneweber, V. Aggressiv sind immer die anderen. *Zeitschrift für Sozialpsychologie,* 1982, *13,* 177–193.

Murray, H.A. *Explorations in personality.* New York: Oxford University Press, 1938.

Olweus, D. Stability of aggressive reaction patterns in males: A review. *Psychological Bulletin,* 1979, *86,* 852–875.

Olweus, D. Familial and temperamental determinants of aggressive behavior in adolescent boys: A causal analysis. *Developmental Psychology,* 1980, *16,* 644–660.

Pepitone, A. The normative basis of aggression: Anger and punitiveness. *Recherches de Psychologie Sociale,* 1981, *3,* 3–17.

Pepitone, A. Introduction to the symposium *Human violence in cross cultural perspective.* International Association for Cross-Cultural Psychology, 6th International Congress, Aberdeen, July 1982.

Weiner, B. *Theories of motivation: From mechanism to cognition.* Chicago: Rand McNally, 1972.

Chapter 3

Individual Differences and Aggressive Interactions

Horst Zumkley

The current status of research on aggression has been increasingly criticized and regarded as unsatisfactory. One argument put forward is that empirical results based on the three traditional approaches – drive theory, frustration-aggression theory, and learning theory – provide only an incomplete, vague, and, in part, contradictory description of the topic. Since continuing on these lines would probably prove nonproductive, new perspectives for aggression research seem necessary.

Social Psychological Perspective: "Aggressive" Is Always the Other Person

The field of social psychology has recently offered some new stimulating approaches (e.g., Tedeschi, Smith, & Brown, 1974; Mummendey, 1980, 1982a). In Germany, Mummendey (1982b), in a comprehensive article, presented the social psychological viewpoint, in which aggressive behavior is regarded as the result of an interaction between individuals and conditions of the social context. Aggression is viewed as the result of an evaluation process in a situation of social interaction; an act is judged by the perceiver (victim) to be aggressive, based primarily on the criteria of intention, harm done, and degree of deviation from the norm.

The basis for the social psychological approach is the analysis of aggressive behavior as a social interaction involving at least two persons (or social units) who find themselves, at some specific point, in a characteristic relationship, that of actor and victim. They are in conflict, in that they *by definition* have incompatible interests, namely, injuring and avoiding injury. Out of this conflict relationship arises a position-specific divergence of judgment about the action directed against the victim with regard to its situational-normative conformity. This acute, simultaneous discrepancy results from the actor's viewing his action as a legitimate, norm-conforming, and justified reaction in the given situation, while the victim regards it as inappropriate and unwarranted. The social context provides norms and rules of social behavior and therefore the standards by which the action is judged. (In the event an action is judged to be aggressive, the primary criteria are harm, intention, and normativeness.) The postulate of evaluation dissent is decisive as a definitional criterion for aggressive behavior, since the positions are reversible in the course of interaction sequences: from the victim's perspective the violation of some norm is observed, justification is demanded, and reactions are provoked which are in turn viewed by the new victim (former actor) as aggression; then a norm-relevant and situationally appropriate reaction is again called for, etc. In other words, it is always the *other* person who is aggressive! According to Mummendey, it follows from the postulate of evaluation dissent that the decision of whether an interaction is to be labeled as aggressive is no longer arbitrary, since foundations for situational and behavioral definitions must incorporate the normative consensus of separate groups. It would therefore seem reasonable that attention be focused on illuminating the surrounding social context which codetermines certain characteristics of aggressive behavior (e. g., norms, social consensus, position, harm, norm deviation, labeling, etc.) (cf. Mummendey, 1981; Mummendey, Linneweber, & Löschper, this volume).

"Aggressive" Is Always the Other Person?: Some Research Gaps

The merit in the aggressive-is-always-the-other-person approach is that it draws obvious attention to the previously neglected multifaceted social structural conditions which codetermine and stabilize aggressive behavior from the outside. The need remains, however, for systematization of individual conditions and – perhaps more important – the definition of the relationship between the social context and the individual. With regard to these two problems, a more motivational psychological view reveals a few points that have been overlooked or disregarded, mainly related to the importance and effect of individual differences.

The social psychological approach described above contains, in my opinion, a view of personality which intrinsically implies that people have a tendency (need/habit) to want to conform to social norms. This is, for example, implicit in the premise that the actor always believes he behaves "appropriately;" he always judges his action positively (otherwise he wouldn't have done it). However, this is not always the case; it applies best in an instance of the (defensive) retaliation norm of equivalent counteraggression (*lex talionis;* negative norm of reciprocity). An imbalance arising in a social relationship is balanced out again, one thing cancels out another, the people are "even" or "quits." What is overlooked is the theoretical possibility (and often existing fact) that (1) the situation-specific aim of wanting to harm others is also activated and (2) individuals can differ with regard to the generalized goal of wanting to harm others. This means that it is quite likely that a varying tendency exists to behave (hostile-)aggressively, spontaneously, and this tendency is dependent upon the situation and/or person; appropriateness and normativeness are brought into play only, for example, to avoid punishment.

In the social psychological approach, individual differences are discussed only through sociological categories, such as position, status, and role, or through the different positions (actor/victim) and their situationally valid norms. Thus, aggressive behavior basically narrows down to two problems: (1) the *re*action of the victim, insofar as it is related to norm and position (and not to individual features) and (2) the problem of escalation versus nonescalation of aggression.

It would therefore seem that the social psychological approach needs to be broadened to include a more systematic regard for individual differences in the form of personality variables or, in the motivational sense, varied, more or less generalized objectives. Although a systematization of individual conditions is recognized to be necessary and there are scattered indications for this [for example, it is said that evaluation of an actor's behavior and the reaction is dependent upon the "social motivation and personal standards" (Mummendey, 1982b, p. 180) of the victim], the itemized way in which the subject is treated, which largely neglects an analysis of more exact functional relationships, shows alone that sociological categories and situationally valid norms are of primary importance. At a key point this becomes very clear: with regard to the question of *how* divergence originates between actor and victim in judging appropriateness, it is assumed "that different *position-specific* emphases and selections of aspects of an 'offered' or 'provided' context give indications for various situational definitions" (Mummendey, 1982b, p. 189). There can be no doubt that this happens situation-specifically, or that – and this will be demonstrated – individual differences also play a decisive role.

It seems to me that consideration of individual differences is especially important with respect to the objective of wanting to harm others – for the actor (spontaneous aggression) as well as for the victim (reactive aggression). In the following I want to demonstrate, using two examples, the importance and mechanism of individual differences in aggression and to show (1) that also in the case of the negative norm of reciprocity there are different standards of retaliation, influenced by a hostile-aggressive objective (goal-orientation); (2) that there is also a varying individual tendency to attribute (hostile) intentions which greatly influences the situational definition and aggressive action; and (3) that in both cases emotional processes (anger, fury) can play a role, the importance of which seems to have been overlooked by the social psychological approach.

Retaliation Norm in Aggressive Interactions and Individual Differences

Aggression is, to a marked degree, subject to social norms. These norms, and therefore the kind and frequency of aggressive behavior, are culture dependent, i.e., each culture has its own norms and evaluations of aggression, and this is where social psychology applies its analysis. The child learns all of these particulars by growing up in his culture during the course of his development. This can be seen in the progressive cognitive and moral development of the child, where a gradual interweaving of moral rules, behavior norms, and aggression occurs. One basic rule, which is learned during the development of moral judgment and retains its validity through adulthood, is the *retaliation norm;* according to Piaget it is typical of the heteronomy stage, which is characterized by the idea of absolutely inviolable rules which one must always follow. Being punished for overstepping one of these rules has an expiatory, effacing quality: one act cancels out another, you're quits. Through personal experiences (being punished), modeling, and peer group experiences, the retaliation norm is further differentiated into balanced counteraggression.

From the viewpoint of motivational psychology, this defensive retaliation norm requires – in addition to determining norm deviation – deciding on an *equivalent* counteraggression (punishment) and this implies: if some act of hostile aggression occurs, then under the retaliation norm, a certain amount of counteraggression must have an optimal satisfaction value (positive incentive of retaliation, of getting even), the motivation for aggression is reduced to zero, and aggressive behavior in similar situations is strengthened. If the counteraggression is too little, there would apparently be no satisfaction, since the hostility wouldn't be completely discharged; on the other hand, too much retaliatory aggression would arouse guilt feelings and/or restitution impulses and could also conceal the danger of renewed counterattacks (negative incentive of re-retaliation). This actually implies – in addition to the evaluation of situational norms – the consideration of retaliation *standards* (as in level of aspiration), which has not yet been realized. The one punished (new victim) must also regard the action as appropriate in order that the desired effect (being even) occurs. The ability to put oneself in another's place (role-taking) is therefore crucial for estimating appropriate counteraggression. In addition to empathy, an even more fundamental attribution process plays a role, that of weighing the other's aggression according to the underlying intention. Both retaliation standards and attribution of motivation appear – as will later be shown – to be influenced systematically by individual differences in aggressiveness (cf. also Berkowitz, 1982; Edmunds & O'Hagan, 1981; Ohbuchi, 1982).

The first example discussed below will show that an "appropriate" retaliation usually does have an optimal satisfaction value, and that if this is too low it does not completely reduce an existing aggressive motivation to retaliate; this retaliation standard can, however, vary according to individual differences in aggressiveness (i.e., the degree of hostile-aggressive goal orientation).

Experiment I

This study was undertaken to discover the effects of a varying amount of retaliatory aggression (varying degrees of success in goal attainment) on aggressive interactions. We investigated the special case in which, after an attack, retaliatory aggression could at first be only indirect, in that the victim was able to complain about the aggressor to his superior and damage his reputation. The degree to which the superior would act on the complaints and actually instigate punishment was varied, i.e., a complete (appropriate), partial, or no retaliation would occur before a new confrontation took place with the aggressor. The investigation (Zumkley, 1978) assessed individual male subjects who were assigned to the experimental groups ($N = 17$ for each group), matched according to scores on the aggressiveness scale of the *Freiburger Persönlichkeitsinventar* [Freiburg Personality Inventory (FPI)]. The experimenter introduced himself as a doctoral student who was collecting data for his dissertation. The attacker/frustrator entered during the experiment and told the experimenter that a Professor Katz had just unexpectedly arrived, had only a few minutes free, and urgently wanted to speak with him. After a short discussion with the experimenter, he used the time while the experimenter was away for his own brief experiment. Following a short test problem, for which all subjects were praised for their good results, came the critical task (counting backwards from 200 in steps of three), for which a reward was offered. During this task the subjects in the frustration groups were interrupted, harshly criticized for their performance, and scornfully informed that they would receive no payment. As the attacker left the room, the experimenter returned. The retaliation (or "catharsis") condition was subdivided into complete retaliation/goal attainment and partial retaliation. In the first case, the experimenter listened sympathetically to the subject's complaints and promised personally to investigate the affair, to make the attacker account for his actions and punish him, and to see that the subject received his money. In the second case, the experimenter listened to the complaints for one minute and then interrupted, saying he would get back to it later and would take care of the money. The attacker then returned, giving the subject another opportunity for a direct confrontation. A third group was given no chance to complain following the provocation (no retaliation), and a fourth group (serving as a control) was not attacked and received the promised reward with no delay.

There were three indicators for state of motivation, measured several times in the course of the experiment: (1) pulse rate as activation indicator; (2) projective tests: (a) form interpretation of a projected, unstructured image (Form Interpretation Test) and (b) Kornadt's (1982) aggression TAT; and mood ratings (anger thermometer). The measures were presented to the subjects as an actual part of the study and were therefore accepted as natural components of the investigation. Pulse rate was measured before the attack (base level) and after the attack (activation) and again following the first opportunity to retaliate (deactivation), or, for the control group, following a neutral activity. Motivation was measured with the TAT at this time, along with the Form Interpretation Test, which had also been given at the beginning. Mood ratings were made after the attack and after the first and second opportunities for retaliation. Spontaneous verbal aggression was recorded directly after the attack and during the second opportunity for retaliation.

The results confirmed our motivation-theoretical expectations, and the most important are illustrated in Figures 3.1, 3.2, and 3.3. Pulse rate, which was the same for all groups at the beginning, increased significantly after the attack, as did the degree of self-reported anger (the control group reported no anger). Corresponding to complete (appropriate), partial, or no retaliation after the first opportunity for it, both excitement level (pulse rate) and experienced anger decreased (see Figure 3.1), depending upon the degree of retaliation (goal attain-

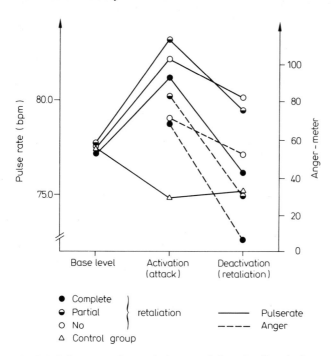

Fig. 3.1. Pulse rate and anger before attack (base level) and after attack (activation) and after different conditions of retaliation (deactivation, i.e., catharsis). (Adapted from Zumkley, 1978)

ment) that was realized. Aggression motivation, measured with Form Interpretation Test and TAT, was not increased in comparison with the control group or base level (cf. Figure 3.2) after complete attainment of the aggression goal (appropriate retaliation). With partial or no retaliation, on the other hand, there was a significant increase in aggression motivation. In the subsequent confrontation with the aggressor, no members of the group with previous appropriate retaliation exhibited direct verbal aggression, nor was there even any mention of the incident (as in the control group; see Figure 3.3). In contrast, the group without retaliation (and to a lesser extent the group with partial retaliation) took advantage of the opportunity for verbal aggressive attacks on the previous attacker.

Taken together, the results show that the intentional injury to one's own interests by another triggers an anger affect which can be reduced by aggressive action (retaliation) against the perpetrator. Expectation of this positive change in affect had a motivating effect. To the extent to which the goal was achieved, a catharsis occurred, i.e., the activated motivation to aggression was deactivated and further aggression against the original perpetrator was reduced. The results also indicate that an appropriate (here a complete) retaliation apparently has an optimal satisfaction value and that the victim's motivation for retaliatory behavior is reduced to zero, while too little or no retaliation cannot fully discharge the resulting hostility, so that aggressive action will be taken up again later. It should

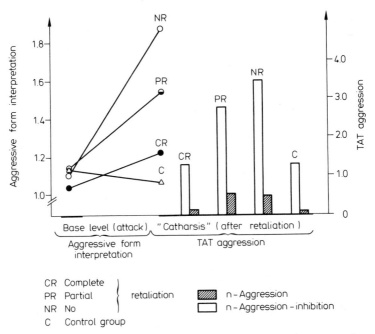

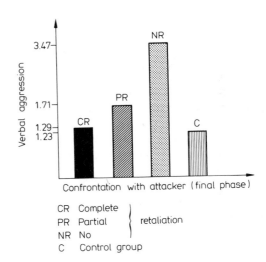

Fig. 3.2. Fantasy aggression (Form Interpretation Test) before attack and fantasy aggression (Form Interpretation Test and TAT) after different conditions of retaliation (deactivation, i.e., catharsis). (Adapted from Zumkley, 1978)

Fig. 3.3. Verbal aggression on confrontation with attacker (final phase of experiment). (Adapted from Zumkley, 1978)

CR Complete ⎫
PR Partial   ⎬ retaliation
NR No   ⎭
C   Control group

also be noted that obtaining the withheld payment cannot have been the primary goal, since the subjects in the group with partial goal fulfillment were certain they would receive their money. Nevertheless, a great decrease or a complete deactivation did not occur and aggressive action was taken up again.

Reanalysis of data with respect to individual differences. As mentioned earlier, the experimental groups were matched according to usual aggressiveness. We then reanalyzed the data in such a way that the groups were divided into high and low aggressive subgroups (high versus low hostile aggression). Results in the subgroups corresponded on the whole (that is, as far as the course of aggressive interactions was concerned) almost exactly with those of the matched groups. There were, however, some important differences. High-aggressive subjects reacted to the attack with significantly more anger ($t = 2.57$; $df = 59$, $p < 0.02$, two tailed). Following complete ($t = 2.17$; $df = 18$, $p < 0.05$) and partial ($t = 3.05$; $df = 18$, $p < 0.01$) retaliation, high-aggressive subjects occasionally showed higher residual anger. A similar picture is given by the fantasy aggression scores. Following complete ($t = 1.73$; $df = 18$, $p < 0.10$) and partial ($t = 2.13$; $df = 18$, $p < 0.02$) retaliation, the aggression motivation was significantly higher for high-aggressive subjects.

We also investigated the particular circumstances in which harm done by the attacker was seen. A characteristic difference was revealed: high-aggressive subjects said they felt they were harmed most "only by the attacker's behavior" and were angered most by this, whereas low-aggressive subjects weighted "behavior and money equally" ($Chi^2 = 9.83$; $df = 1$, $p < .01$). In addition, when there was no promise of retaliation (punishment), high-aggressive subjects tended to show significantly more spontaneous verbal aggression towards the attacker in the second confrontation with him than did the low-aggressive subjects ($t = 1.72$; $df = 20$, $p < .10$).

Without putting too much emphasis on this result, it can nevertheless be said that individual differences in aggressiveness above and beyond normative conditions are apparently important in aggressive interactions. Aggressive people react differently to the same provocation, with more anger, and following complete (appropriate) as well as partial retaliation they are less satisfied than low-aggressive persons under the same conditions. This stronger anger caused by the attack, which also produces a stronger aggression motivation, might reflect the subjective feeling of a stronger norm violation than is actually the case, which in turn requires stronger retaliation; in our experiment, for example, this was expressed by greater spontaneous verbal aggression in the new opportunity for aggressive interaction. One could also interpret this by saying that a goal of harmful intent rose above and beyond getting even, which works as a coactivator in high-aggressive people and which must also be satisfied.

Individual differences in aggressiveness seem to be related to the number of anger-related cues (extensity) and to the intensity of the emotional reaction (stronger aggression motivation) which involves higher retaliation standards in the sense of individual reference norms and/or others, namely hostile-aggressive goal expectancies for retaliation in highly aggressive persons. This also conforms with Edmunds and O'Hagan's (1981) finding that highly aggressive people in an aggression-relevant situation seem even to start out with other reference and, therefore, retaliation norms for aggressive interaction. The authors were able to demonstrate that over different groups of subjects global judgments of aggressive interactions are fairly uniform, but that systematically varying divergences of judgment arise according to individual differences; these permit a stronger acceptance of aggression in general; they rate hostile-aggressive forms of aggressive interactions significantly higher (i.e., physical attacks also against innocent victims) and use these forms in their own behavior – spontaneously as well as reactively – more often.

Causal Ambiguity in Aggressive Interactions and Individual Differences

An action is primarily judged to be aggressive if intent to harm is perceived. This motivation attribution also is decisive in determining the course of aggressive interactions. The following two examples will show how certain situational contexts regularly guide the perception of motivation attributions, the effects these usually have on the course of aggressive interactions, and how individual differences in aggressiveness codetermine such situational judgments and the course of the interactions.

According to attribution theory, the strength of a retaliatory aggression is very dependent upon whether the attack is regarded as a personal desire of the attacker, and thereby intentional and dispositionally determined (internal cause), or whether it is unintentional, caused by situational circumstances (external cause) (Jones & Davis, 1965; Kelley, 1971, 1973). While both types of causes are considered and weighed for the retaliation, according to these results hostile intentions tend to be attributed to the attacker more often when fewer external causes are present; angry emotion and retaliatory aggression are then correspondingly higher. This can be related to Kelley's "discounting principle" and Deci's (1975) specification of it: in explaining actions and their results, internal causes are sought out when external causes do not provide sufficient explanation. In the event that several plausible causes exist, attribution is diminished (devalued), i.e., becomes less extreme or less unequivocal than when only one plausible cause is present. According to Deci "this implies that if a behavior occurs in the presence of a facilitating environmental force, *internal* factors will be discounted" (Deci, 1975, p. 247). Enzle, Hansen, and Lowe (1975) found that this precedence of external over internal causes holds true in general: when personal as well as circumstantial causes are available as plausible explanations so that *causal ambiguity* exists, then attribution to external causes has priority.

Since Enzle et al. (1975) studied only cooperative and competitive behavior exclusively with female subjects, the first purpose of our investigation was to study the validity of this specification of the discounting principle as applied to aggressive interactions with both male and female subjects; sex differences in aggression have been confirmed many times (Frodi, Macaulay, & Thome, 1977; Merz, 1979; Rothmund, 1979). We were also interested in an aspect of information processing in connection with attribution, the question of observing and recording various bits of causal relevant information at different times in the course of aggressive interaction sequences, and their effect on retaliatory aggression and physiological arousal – a relatively unexplained process up to now (cf. Kelley & Michela, 1980, p. 472).

Experiment II

These considerations required an experimental arrangement of conditions which would make possible the transmission of information before or during the provocation phase (attack) – following Kelley's model (see Table 3.1) – about low consensus, low distinctiveness, and high consistency, so that a personal attribution would be highly likely.[1] In the dependent variables we expected effects in the form of attributions of greater hostile intention, stronger internal direction, and greater responsibility for the attack behavior, together with strong subjective anger, high physiological activation, and strong behavioral aggression in the victim. Another group of subjects was studied to show whether – also in the case of an obvious internal cause – subsequent information about high distinctiveness of the attack with the suggestion of competing and variable external causes (i. e., situational attribution) actually results in a discounting of internal causes, as would be expected from Deci and Enzle. If this proved to be the case, then a change in attribution would have to occur, i. e., attribution of harmless intent together with a reduction in anger and less retaliatory aggression. A $2 \times 2 \times 2$ design was used with sex of subjects, sex of attacker, and low/high distinctiveness as independent variables; for the last-named independent variable, in one case no more information given after the attack, in the other some information about high distinctiveness of the attacker's behavior (hostile intent versus ambiguous intent condition).

Table 3.1. Information patterns leading to entity, person, or circumstances attributions (cf. Kelley, 1971, 1973)

Attribution	Consensus	Information distinctiveness	Consistency
Entity	High	High	High
Person	Low	Low	High
Circumstances	Low	High	Low

Ninety-six students (of which 48 were female) took part in the experiment (Zumkley, 1982a). The separate parts of the study followed the basic pattern of a procedure by Zillmann and Cantor (1976), which we had used earlier (Zumkley, 1981). It consisted of an interaction between a subject and two experimenters, one of whom is courteous and correct (E = experimenter), the other rude to aggressive (A = attacker). In the experiment the subject observed how A unjustly, rudely, and aggressively behaved toward E, once before and once after the provocation phase (attack on subject). This gave E a plausible opportunity during the

---

1 Only information about the distinctiveness of the attack could be varied here (low/high) with low consensus being present simultaneously under both conditions. The question of the consistency of the attacker's behavior toward the subject at other times had to remain open for the time being; the experimental procedure should make interpretation of both high and low consistency possible.

experiment to impart information on consensus and distinctiveness. The experiment was presented to the subject as a psychophysiological perception study. The subjects were told that their task was to interpret visual stimuli (films and pictures) varying in structuredness, during which their pulse rate was measured. The experiment began with the remark that sitting still and remaining calm were necessary for correct measurements, and this was followed by an association drill and a task which consisted of silently counting irregular taps. This was followed by Film 1 with unstructured content which the subject had to interpret, then the counting task again. During this an interaction took place between the experimenters in which A accused E of having hooked up the physiological apparatus incorrectly. When this was shown not to be the case, A continued to complain, at which point E gave all groups the information on low consensus: "For heaven's sake, pull yourself together, we're in the middle of an experiment. You should behave better, if you're going to work as a practicum assistant." A then left without apology, and E delivered the information on low distinctiveness: "He's always the same, every chance he gets he's rude to other people, even when there's no reason." In a short time, A called E out of the room, saying he was wanted urgently and took over the experiment himself. During this part of the experiment A unjustly attacked the subject verbally; he yelled that the subject was spoiling the measurements by not sitting still and not trying. Then Film 2 was shown and again interpretations were requested. When E returned, A yelled that E hadn't prepared the slides for the second part of the experiment, but again this proved not to be true; A then left the room without apologizing. Following this, one of the groups received no further information, the other group the following information on high distinctiveness: "I completely forgot that he has final exams tomorrow – I can understand why he's so distracted today." Then Kornadt's TAT was performed, following which the subjects completed the Adjective Checklist (ACL) by Janke and Debus (1978) and two questionnaires in another room with a third experimenter.

Dependent variables included: pulse rate as an indicator of physiological arousal (activation), measured immediately after the attack (t_1), immediately following the information on high distinctiveness (t_2), and at a comparable time in the groups which received no further information. Complaints and retaliatory aggression were recorded in a confidential questionnaire and the subjective reactions in the ACL. Fantasy aggression was measured by analyzing the interpretations of the unstructured film materials according to Hafner and Kaplan's scale (1960) and by using Kornadt's (1982) method to score the TAT for aggression and aggression inhibition. A third questionnaire was used for recording causal attributions. The results are summarized in Table 3.2. They confirm the theoretical postulate from Kelley's model regarding the necessary intensity and combination of the information components of the consensus (low) and distinctiveness (low) for personal attribution in the case of aggressive behavior. The attack was carried out as if with strong hostile intention and judged to be internally guided. This resulted in a strongly pronounced anger and an aggression-related change in the projective behavior of form interpretation (Form Interpretation Test). This also produced a behavioral change in the form of aggressively toned complaints

Table 3.2. Mean differences among the eight experimental groups in a hostile intent and an ambiguous intent condition

Group / Dependent variables	I Male subjects Male attacker Low distinctiveness	II Male subjects Male attacker High distinctiveness	III Male subjects Female attacker Low distinctiveness	IV Male subjects Female attacker High distinctiveness	V Female subjects Male attacker Low distinctiveness	VI Female subjects Male attacker High distinctiveness	VII Female subjects Female attacker Low distinctiveness	VIII Female subjects Female attacker High distinctiveness
Hostile vs harmless intention	55.42[a]	47.08[ab]	40.83[ab]	-8.33[d]	24.17[bc]	-5.42[d]	26.25[b]	3.75[cd]
Internal vs external causality	32.92[a]	26.67[a]	25.42[a]	-0.83[b]	12.92[ab]	-2.50	11.25[ab]	7.92[b]
Personal vs external responsibility	8.75[a]	4.58[a]	6.25[a]	2.92[b]	5.83[a]	-1.25[a]	5.00[a]	3.33[a]
Pulse rate	-0.58[abc]	-0.33[abc]	-0.75[abc]	-3.33[*]	1.17[ab]	-3.17[*]	3.42[*]	0.67[abc]
Fantasy aggression (form interpretation)	2.00[a]	2.17[a]	1.67[a]	1.92[a]	1.75[a]	2.25[a]	2.00[a]	1.83[a]
TAT-aggression	1.25[a]	1.08[a]	0.92[a]	1.17[a]	1.08[a]	0.75[a]	1.00[a]	1.08[a]
TAT-aggression inhibition	0.50[a]	0.42[a]	0.42[a]	0.58[a]	0.75[a]	0.50[a]	0.92[a]	0.67[a]
ACL-anger	4.50[a]	2.08[b]	3.92[a]	1.92[b]	3.83[a]	1.75[b]	4.17[a]	1.83[b]
ACL-activation	7.67[bcd]	5.58[d]	6.58[d]	6.92[d]	11.08[ab]	10.42[abc]	11.58[a]	11.33[a]
ACL-anxiety	1.67[b]	1.83[b]	2.08[ab]	1.50[b]	4.17[a]	3.08[ab]	3.17[ab]	2.75[ab]
ACL-self-reliance	2.00[b]	4.33[a]	1.67[b]	1.83[b]	1.58[b]	1.67[b]	2.00[b]	1.92[b]
ACL-depression	5.25[a]	5.42[a]	6.17[a]	5.67[a]	9.25[a]	5.92[a]	6.33[a]	5.75[a]
General dissatisfaction	48.75[abc]	22.92[c]	44.58[abc]	28.33[bc]	55.00[ab]	34.17[bc]	67.50[a]	41.67[abc]
Excessiveness of requirements	17.08[a]	14.58[a]	15.42[a]	14.17[a]	15.00[a]	15.83[a]	20.42[a]	19.17[a]
Performance of attacker	-45.42[b]	-36.67[b]	-31.25[b]	23.33[a]	-24.58[b]	26.67[a]	-47.08[b]	-5.42[ab]
Courtesy of attacker	-55.00[c]	-22.92[abc]	-45.83[c]	8.33[a]	-36.67[bc]	-8.75[ab]	-51.25[c]	-32.08[bc]
Reappointment of attacker	-42.08[d]	-36.25[d]	-18.33[cd]	16.25[ab]	-18.75[cd]	25.42[a]	-26.67[cd]	-6.67[bc]
Performance of experimenter	50.00[a]	55.00[a]	55.42[a]	52.92[a]	51.67[a]	49.58[a]	47.50[a]	53.33[a]
Courtesy of experimenter	68.33[a]	65.00[a]	67.08[a]	62.08[a]	72.92[a]	65.42[a]	66.25[a]	69.17[a]
Reappointment of experimenter	60.83[a]	63.33[a]	60.00[a]	59.58[a]	58.75[a]	56.25[a]	64.17[a]	65.42[a]

Note. Letter superscripts denote horizontal comparisons. Means having a superscript letter in common do not differ significantly ($p < .05$) by the Newman-Keuls test (cf. Sachs, 1974, p.410f.). Pulse rate scores are relative to basal measure in bpm ($t_1 - t_2$). Means denoted by asterisk are significantly different ($p < .02$) from base level (0) by t test (cf. Sachs, 1974, p.201f.). (Based on data from Zumkley, 1982a).

and a strong retaliatory aggression directed solely at the original attacker. Sex differences on the whole remained minimal. Subsequent information on the high distinctiveness of the attack behavior caused a discounting of the internal cause in the corresponding groups: an attribution change occured in the form of assigning harmless, nonhostile intention and less internal guiding of behavior, as well as a change in affect (anger reduction and physiological deactivation), which in turn produced an alteration in coping with the attack, in the form of fewer complaints and less retaliatory aggression. There was only one exception: in the aggressive interaction between males (Group II, cf. Table 3.2), this attribution change did not occur, and the changes in the dependent variables weren't all in the expected direction. We presume this was due to a sex-specific difference in credibility.

On the whole, therefore, our results conform with Deci and Enzle's specification of Kelley's discounting principle, which states that internal causes are used to explain actions and their consequences only insofar as external causes are not sufficient or do not exist; external causes do indeed appear to take precedence.

Experiment III

The inconsistency in the results for aggressive interactions between male partners induced us to repeat the experiment. We wondered whether our assumption of a sex-specific difference in credibility could arise due to individual differences

Table 3.3. Mean differences among subjects with high and low n-aggression scores in an ambiguous intent condition

Dependent variables	n-Aggression		$p<$ [a]
	High	Low	
Hostile vs harmless intention	69.74	−5.00	0.001
Internal vs external causality	34.47	−27.63	0.001
Personal vs external responsibility	1.32	−3.42	n.s.
Pulse rate[b]	0.79	−2.21	0.01
Fantasy aggression (form-interpretation)	2.16	1.84	0.10
Fantasy aggression (TAT)	1.53	0.74	0.10
ACL-anger	4.47	1.95	0.001
ACL-activation	9.58	7.95	0.05
ACL-anxiety	1.59	1.47	n.s.
ACL-self-reliance	3.68	4.00	n.s.
ACL-depression	6.42	6.63	n.s.
General dissatisfaction	61.58	27.89	0.001
Excessiveness of requirements	35.00	13.68	0.001
Performance of attacker	−42.68	−5.00	0.002
Courtesy of attacker	−68.42	9.21	0.001
Reappointment of attacker	−42.89	21.05	0.001
Performance of experimenter	53.42	48.68	n.s.
Courtesy of experimenter	58.42	57.63	n.s.
Reappointment of experimenter	44.21	51.58	n.s.

[a] t-test, two tailed; df $= 36$. [b] Pulse rate scores are relative to basal measure (t_1–t_2).

in a specific person-situation interaction. We studied (Zumkley, 1982b) male subjects divided into groups of high and low aggressiveness based on usual aggressiveness (aggression-motive strength on Kornadt's (1982) TAT). Both groups received the ambiguous intent condition; that is, the subjects first received the information on low consensus and low distinctiveness and following the attack, information on high distinctiveness of the behavior. The procedure followed was that described in Experiment II.[2]

The results are summarized in Table 3.3. Characteristic differences depending on individual differences in aggressiveness can be seen: high-aggressive subjects apparently disregarded the additional mitigating information and did not devalue the internal cause. On the contrary, they demonstrated – across all groups – the highest hostile intention attribution, combined with attribution of strong internal behavioral control. Increased pulse rate caused by the attack remained high – only in low-aggressive subjects did deactivation occur – and there were clear indications for higher aggression motivation (fantasy aggression). This was at the same time combined with significantly greater anger and excitement, as evidenced by the pulse rate, remaining high; in low-aggressive subjects these both decreased following attribution change. In addition, high-aggressive subjects showed a highly significant greater dissatisfaction with the experiment and were the only who judged the requirements of the experiment (otherwise regarded as minimal) to be excessive. With regard to retaliatory aggression against the attacker, they showed, on all counts, higher retaliatory aggression than low-aggressive subjects; the courteous experimenter, however, was rated equally favorably by both groups.

Our results are in accordance with those obtained by Dodge (1980). In his experiment with children in grades 2, 4, and 6, aggressive and nonaggressive subjects also differed in an ambiguous intent condition: aggressive boys responded to ambiguous-intention, negative-consequence situations with strong aggression against a peer, whereas nonaggressive subjects responded as if the peer had acted with a benign intent. Taken as a whole, our results also confirm a general trend for precedence of external over internal causes in the usual case. Individual differences in aggressiveness seem, however, to cause a specific person-situation interaction and to play a decisive role as antecedents of motivational attributions in connection with aggressive interactions. This apparently implies different causal beliefs, i.e., general, preexisting (learned) assumptions (causal schemata) about causes for certain kinds of events and – for any given cause –

---

2 The subjects were grouped according to aggression and aggression inhibition scores on Kornadt's (1982) Aggression TAT (K-TAT). Classification criteria used were: (1) high-aggressive subjects: above-average aggression score and below-average or average aggression inhibition score, based on norms provided by Kornadt (1982, Vol.2, pp.89ff.); (2) low-aggressive subjects: below-average aggression score and below-average or average aggression inhibition score. The groups ($N=19$ in each case) only differ, therefore, in TAT aggression scores, not in TAT aggression inhibition scores. Since the groups were divided according to K-TAT scores, in Experiment III (as opposed to Experiment II) two pictures from the Murray TAT were used, cards 1 and 8 BM, selected for their scaled values for aggression in accordance with Murstein (1963, p.220). The stories for these two pictures (as well as the films) were scored for fantasy aggression according to the Hafner and Kaplan Scale (1960).

expectations about their consequences. According to our results, such preexisting convictions actually seem to influence the acquisition and evaluation of new, causally relevant information as well (cf. Kelley & Michela, 1980, p.473). One could also see in this – in agreement with Ohbuchi's (1982) results – the effectiveness of an attribution bias and assume that highly-aggressive persons tend toward negativity bias in motivation attributions, i.e., that individual differences occur and these people give more weight to negatively valenced information (here low consensus and low distinctiveness) in intention attribution. In an event, in the normal case subsequent information about the possibility of an external cause for aggressive behavior resulted in a more rational processing of it, and therefore to a devaluation of internal cause and a corresponding change in retaliatory behavior. This did not happen in high-aggressive people; their reaction was rigid. The intention attributions called into being and their presumably related causal convictions remained highly emotional (strong anger, high arousal) and resulted in correspondingly stronger retaliatory behavior.

Concluding Remarks

From the motivation-psychological standpoint, an individual in an aggressive interaction is an active actor, following intentions, and he necessarily attracts a larger share of investigative attention than the (central from the social psychological standpoint) context conditions, which we also took into account here. Present behavior is regarded as a function of a continuous process of a multifaceted interaction. This interaction concept is different from the one which forms the center of the social psychological analysis (cf. Mummendey et al., this volume); here interaction represents a feedback between the individual and the situation he enters. For this reason the most important determinants of behavior on the individual's side of the interaction are cognitive and motivational factors, whereas on the situation side the psychological relevance a situation has for the individual is decisive (cf. Magnusson & Endler, 1977, p.4).

According to this model, individuals react, but not only to situations (as strongly implied by the social psychological approach); they also select situations according to their dispositions and/or create situations. They therefore limit the number of possible situational influences, and select and accent certain aspects, thereby producing a situational specifity. My purpose was to demonstrate that this is a determinant of aggressive behavior, relatively overlooked in the social psychological perspective of aggression research, for which individual differences in aggressiveness are responsible (i.e., motivational differences in the form of varying degrees of the general goal orientation of wanting to harm others). Thus, aggressive behavior cannot simply be described according to individual variations in degree of motive strengths, but rather according to their varying efficiency in different situations and their interpretation. Aggressive behavior is neither "caused" by a dominating motive nor is it "controlled" by the situation. Both are dependent on each other in effecting behavior.

An impressive amount of evidence has demonstrated the stability, i. e., cross-situational and longitudinal consistency of individual differences in aggressive reaction patterns (motive systems) in individuals (see Olweus, 1979; Loeber, 1982). Individual differences in aggressiveness which decisively influence spontaneous and reactive aggressive behavior occur, as we have shown, as a result of, for example, the number of anger-related cues (extensity), the intensity of the emotional reaction, and the tendency toward hostile motivation attributions which apparently arise due to differences in causal convictions (causal schemata) and which appear to influence the processing of information as well. These variables obviously seem to modify at least some of the basic processes necessary to account for aggression. Moreover, a motivation psychological perspective on aggression (Kornadt, this volume; see also Olweus, 1978) offers a rationale for studying the individual difference variable systematically and not in a random manner. Seen in this light, individual differences must also be important for the social psychological approach, since the evaluation process in a social interaction situation (a decisive and fundamental cornerstone of the theory) is systematically codetermined by them.

References

Berkowitz, L. Violence and rule-following behavior. In P. March & A. Campbell (Eds.), *Aggression and violence*. Oxford: Blackwell Publications, 1982, 90–101.

Deci, E. *Intrinsic motivation*. New York: Plenum Press, 1975.

Dodge, K. Social cognition and children's aggressive behavior. *Child Development, 1980, 51,* 162–170.

Edmunds, G., & O'Hagan, F. Evaluation of different types of aggressive acts among a variety of male adolescent samples. *Journal of Adolescence, 1981, 4,* 249–259.

Enzle, M., Hansen, R., & Lowe, C. Causal attribution in the mixed-motive game: Effects of facilitory and inhibitory forces. *Journal of Personality and Social Psychology, 1975, 31,* 50–54.

Frodi, A., Macaulay, J., & Thome, P. Are women always less aggressive than men? A review of the experimental literature. *Psychological Bulletin, 1977, 84,* 634–660.

Hafner, A., & Kaplan, A. Hostility content analysis of the Rorschach and the TAT. *Journal of Projective Techniques, 1960, 24,* 137–143.

Janke, W., & Debus, G. *Die Eigenschaftswörterliste (EWL)*. Göttingen: Hogrefe, 1978.

Jones, E., & Davis, K. From acts to dispositions. In L. Berkowitz (Ed.), *Advances in experimental social psychology* (Vol. 2). New York: Academic Press, 1965, 219–266.

Kelley, H. Attribution in social interaction. In E. Jones (Ed.), *Attribution: Perceiving the causes of behavior*. Morristown, N.J.: General Learning Press, 1971, 1–26.

Kelley, H. The process of causal attribution. *American Psychologist, 1973, 28,* 107–128.

Kelley, H., & Michela, J. Attribution theory and research. *Annual Review of Psychology, 1980, 31,* 457–501.

Kornadt, H.-J. *Aggressions-Motiv und Aggressions-Hemmung*. (2 Bände). Bern: Huber-Verlag, 1982.

Loeber, R. The stability of anti-social and delinquent child behavior: A review. *Child Development, 1982, 53,* 1431–1446.

Magnusson, D., & Endler, N. (Eds.), *Personality at the cross-roads: Current issues in interactional psychology*. Hillsdale, N.J.: Erlbaum, 1977.

Merz, F. *Geschlechtsunterschiede und ihre Entwicklung*. Göttingen: Hogrefe, 1979.

Mummendey, A. *When are persons willing to compensate their victims?* Paper presented at East-West Meeting of the European Association of Experimental Social Psychology at Bologna, Italy, May 13–16, 1980.

Mummendey, A. *Aggressive Interaktionen in der Schule*. Unveröffentlichter Forschungsbericht II. Psychologisches Institut der Universität Münster/Westf., 1981.

Mummendey, A. *Zum Nutzen des Aggressionsbegriffs für die psychologische Aggressionsforschung*. In R. Hilke & W. Kempf (Eds.), *Aggression: Naturwissenschaftliche und kulturwissenschaftliche Perspektiven der Aggressionsforschung*. Bern: Huber-Verlag, 1982, 317–333. (a).

Mummendey, A. Aggressiv sind immer die anderen – Plädoyer für eine sozialpsychologische Perspektive in der Aggressionsforschung. *Zeitschrift für Sozialpsychologie*, 1982, *13*, 177–193. (b).

Murstein, B. *Theory and research in projective techniques (emphasizing the TAT)*. New York: Wiley, 1963.

Ohbuchi, K. Negativity bias: Its effects in attribution, hostility, and attack-instigated aggression. *Personality and Social Psychology Bulletin*, 1982, *8*, 249–259.

Olweus, D. *Aggression in the schools: Bullies and whipping boys*. Washington, D.C.: Hemisphere, 1978.

Olweus, D. Stability of aggressive reaction patterns in males: A review. *Psychological Bulletin*, 1979, *86*, 852–875.

Rothmund, H. Geschlechtsunterschiede im aggressiven Verhalten. In H. Keller (Ed.), *Geschlechtsunterschiede*. Weinheim: Beltz-Verlag, 1979, 75–92.

Sachs, L. *Angewandte Statistik*. Berlin: Springer-Verlag, 1974.

Tedeschi, J., Smith, R., & Brown, R. A reinterpretation of research on aggression. *Psychological Bulletin*, 1974, *81*, 540–562.

Zillmann, D., and Cantor, J. Effects of timing of information about mitigating circumstances on emotional responses to provocation and retaliatory behavior. *Journal of Experimental Social Psychology*, 1976, *12*, 38–55.

Zumkley, H. *Aggression und Katharsis*. Göttingen: Hogrefe, 1978.

Zumkley, H. Der Einfluß unterschiedlicher Absichtsattributionen auf das Aggressionsverhalten und die Aktivierung. *Psychologische Beiträge*, 1981, *23*, 115–128.

Zumkley, H. Kausale Ambiguität: Vorrangigkeit externaler Ursachen? *Psychologische Beiträge*, 1982, *24*, 224–241. (a)

Zumkley, H. Einflüsse von Motiv-Differenzen der Aggressivität bei kausaler Ambiguität. Unpublished manuscript, University of the Saar, Saarbrücken, 1982. (b)

Chapter 4

Aggression as Discourse

Kenneth J. Gergen

To speak is simultaneously to engage in world construction. This is so in two major senses, the first implicative and the second pragmatic. In the former sense, world construction depends on the fact that language serves not as an arrangement of sounds, but as a system of symbols. To qualify as symbols, linguistic entities must imply a realm of referents; not to do so would be to lose identity as language. Thus to engage in the production of language is typically to forge an implicit commitment to a realm of referents not contained within the language itself. In effect, the act of speaking invites the listener to accept an independent ontological system. At the same time, language also has a pragmatic, or what Austin (1962a) has called a performative, aspect. That is, it is much like moving a rook in chess, holding someone in an embrace, or looking disinterested while under attack. It is itself a form of social intercourse. And, as performance, it frequently has social effects. Depending on the language one employs, another may admit defeat, profess deep love, or even kill. One's words are thus active constituents in a world of ongoing social interchange (Searle, 1970).

That language does imply an ontology, or serve to objectify, is not particularly controversial, so long as it is possible to establish linkages between the system of symbols and a range of relevant particulars. Typically such linkages are established through a process of ostensive definition, that is, through demonstrating or setting the experiential context for word usage. In using the word "cat", the mother may point to a small furry object with four legs and thereby establish the referent for the term; as the term is used in the presence of a variety of furry creatures over time, the child essentially learns that the term can refer to a range of objects. As Quine (1960) has made clear, the precise qualities of the referents may remain obscure; the referents for the term may essentially comprise a "fuzzy set" (in this case both literally and figuratively). Yet, through the process of ostensive identification, the world created through language may be treated, for all practical purposes, as possessing a reality independent of the language itself. By the same token, it is this process of ostensive definition that furnishes the major basis for the practical deployment of the language. Such phrases as, "You have jam on your nose," "Your check has arrived," or "Your house is on fire," have important consequences, primarily because of the rough state of affairs that they symbolize.

For purposes of clarity those linguistic descriptors linked to the realm of immediate observables (for some persons at some time) can be said to rely on *direct ostensive grounding*.[1] Such terms may be contrasted with two additional types of descriptors, the first of special importance to much natural science inquiry, and the second, as we shall see, of paramount importance to the understanding of aggression. Those descriptors subject to *indirect ostensive grounding* gain their immediate legitimacy through their linkages to other descriptors; however, this secondary tier of descriptors is itself subject to direct ostensive grounding. Thus, for example, the term "gravity" does not refer directly to any event or set of events in particular. However, the term is linked to sets of descriptors that are themselves ostensively grounded. To say that "the ball falls to the earth" may through common practice be tied to a fuzzy set of observable events (widespread agreement can generally be obtained as to whether a ball is or is not falling to the ground). To say that the ball's movement is the "result of gravitational force" is to establish a linguistic context for the usage of the term "gravity." When one is permitted by ostensive definition to say "the ball is falling," one also has warrant to speak of gravity. Descriptors subject to indirect ostensive grounding may prove problematic when the common linking practices (either from the indirect to the direct descriptor or from the latter to the realm of experience) are either too flexible or conflicting. However, with sufficient negotiation, it should be possible, in principle, to reach general agreement that indirect descriptors could be challenged or corrected by observation. If dropped objects suddenly begin to "ascend into space," the common theory of gravity might well be subject to emendation.

Yet, there is a third class of descriptors that are more problematic in character. The grounding of these terms is based neither on direct nor ostensive grounding, but on equivalence functions within the language itself. That is, the descriptive terms are legitimated through reference to other linguistic integers. In effect, they are *linguistically grounded*. This grounding essentially depends on a system of equivalency functions, or rules that determine the conditions under which the descriptors may appropriately be used. The rules essentially indicate a range of words that, together, would function as the equivalent of the descriptor in question. In the most primative case, these integers may simply serve as synonyms of the descriptor. Consider, for example, the descriptor "obedient," as in "Richard is an obedient man." One is hard put to locate a set of spatiotemporal constituents of the term "obedient"; the term simply does not derive its warrant for usage from any particular arrangement of the muscles, skeleton, neurons, and so on. Richard may presumably be defined as obedient regardless of the displacement of limbs, facial muscles, etc. However, the term is made intelligible because it is tied through equivalence rules to other integers in the language. If one is asked

---

1 The process by which words become defined through the referencing of observables is extensively described by Rommetveit (1968).

what is meant by the sentence, "Richard is an obedient man," for many purposes it would be sufficient to answer that he obeys orders, follows the rules, and avoids displays of autonomy. Fundamentally, such terms are the functional equivalents of obedience; if asked to produce a synonym for these terms, obedience would be a prime candidate. The equivalency system is essentially closed with regard to external or "real-world" grounding. In the same way, the term "heaven" gains legitimacy. Although the term itself has no spatiotemporal coordinates of common access, pervasive equivalency rules enable us to understand it as "the dwelling place of God." The latter phrase is similarly without referential exit to observables, but the user of the phrase may be successfully understood as speaking about "heaven."

This third class of linguistically grounded descriptors is both as problematic as it is essential within social life.[2] Such descriptors are problematic because they carry with them all the pragmatic force of the preceding classes, but without the same warrant. For example, in the case of ostensive grounding, one may warn, "You should leave at once; the building is on fire," and, if the advice is not heeded, the consequences may be lethal. The pragmatic consequences of the message are independent of the language itself. Employing the same linguistic form one might warn, "You should leave this place at once or you will go to hell." This utterance may carry with it the same pragmatic force as the warning of fire. Yet one is hard put in this case to link the travels to hell with a range of observables. The term "hell" is treated as an ontological equivalent to "fire." Although the latter is no more or less "real" than the former, the latter term can be linked to observables in a way that often has practical advantages. The term "hell" can only be grounded in other language. Its pragmatic power is essentially borrowed.

In general, it may be proposed that the social effects ("perlocutionary," in Austin's terms) of descriptive terms is derived in major part from the extent to which the ontological system implied by the descriptors is linked in a practical way with ongoing events. If the language can be ostensively defined, either directly or indirectly, such definitions can furnish the basis for informed and adaptive action. Terms with direct and indirect ostensive grounding should thus have considerable power in daily affairs. Yet, utterances such as, "You will go to hell if you don't . . .," and "Richard is very unhappy today," may also have important social consequences, and these utterances cannot be ostensively grounded. In some degree the power of such linguistically grounded descriptors may be derived from their formal similarity to the ostensive descriptors. Yet, this dependency must be viewed as partial. It is not that the threat of hell has lost its force in much of contemporary society because of the slowly emerging realization that its warrant is solely linguistic. Rather, much of the pragmatic force of such descriptors is derived from common practices of social sanctioning. In many sectors of society a man who is told that his conduct will take him to hell is wise to change

---

2 Elsewhere (Gergen, 1982) I have tried to demonstrate that most descriptions of human action are linguistically, and not empirically, grounded and that, given the particular character of the denotative process, it could scarcely be otherwise.

his ways, not because of this particular likelihood, but because of the social dis-approval he is otherwise likely to incur. In Pennsylvania Dutch culture, a man who engages in "sinful" activity may be systematically shunned by an entire community; no one in the community may be allowed to speak with him or in any other way acknowledge his existence. Similarly, if a friend tells you her mother has just died and you announce that you are feeling happy about it, her dismay will not be the result of an ontological breach, but a social one. It is a breach in the social rules of emotional utterances that may result in the broken friendship. In effect, linguistically grounded descriptors may often serve impor-tant social functions, and their warrant and significance in the culture should properly be traced to these functional bases.

Aggression as Linguistically Grounded

Common discourse on aggression is world constructing in both of the previously described modes. Sentences such as, "The Russians are guilty of aggression in Afghanistan," "American culture is highly aggressive," "Patterns of human ag-gression are similar to those in various animal communities," and "There is a positive relationship between ambient temperature and aggression," first imply an ontology. It follows from such utterances that included in the inventory of en-tities making up the world are various forms of behavior, one of which is aggres-sion. Certain people at certain times display this behavior, and such behavior is different in its fundamental character from behaviors which are not aggressive. Such utterances also possess a variety of pragmatic implications. To say that the Russians or the Americans display aggression, for example, is to have different implications for how they are to be treated than if their behavior under such cir-cumstances were described as "self-protective," "energetic," or "idealistic." The pragmatic interests served by the scientific use of the term aggression have been explored elsewhere by Lubek (1979).

The uses to which the term aggression is put in the above instances might pass without interest if the term were subject either to direct or indirect ostensive grounding. If such grounding were accomplished, then to say that the Americans are aggressive people would have much the same ontological status as saying, "There is no snow on Mount Fuji." Spatiotemporal reorderings should be possi-ble in both instances. In the case of Mount Fuji, independent observers should all be able to reach agreement regarding the existence of snow; photographic re-cordings might also be employed, or inferences could be made from various in-dicators of temperature and precipitation. This is not to presume the indepen-dent existence of snow, but it is the case that by virtue of observation, broad agreement can easily be reached regarding the relevance or adequacy of the term snow. Yet, what are the spatiotemporal exemplars of the term aggression? At what velocity should the body be moving? At what angle should the left femur be extended? What adjustments must be made by stretch reflexes, and at what rate should sodium ions flow into the neurons for an action to qualify as ag-gression? All such questions are without answers, it would appear, because the term aggression is not one of those descriptors linked ostensively to ongoing events.

If the term aggression does not refer to a range of spatiotemporal particulars, then to what does it refer? Perhaps the most adequate answer to this question is furnished by those most deeply invested in developing behavioral indicators of aggression, namely behavioral scientists. Interestingly this issue has been fraught with conflict since the concerted study of aggression began in the 1930s. In their groundbreaking treatise, Dollard, Doob, Miller, Mowrer, and Sears (1939) defined aggression as "an act whose goal-response is injury to an organism" (p. 11). Yet, although it was wrought within the behaviorist framework, investigators were quick to see that this definition was, in fact, nonbehavioral. That is, its ultimate reference was the "goal-response," an internal or psychological construct used to earmark the directionality of behavior. To correct this state of affairs, Buss (1961) later proposed to define aggression as "a response that delivers noxious stimuli to another organism" (p. 1). However, this definition drew extensive fire (cf. Bandura and Walters, 1963; Feshbach, 1964; Kaufmann, 1970), in large measure because it failed to take account of the motive of the actor. As it was argued, Buss' definition would allow the actions of a surgeon or a dentist to be termed aggressive, as would any accidental act resulting in another's pain or death. In light of these criticisms, Buss later altered his definition to "the attempt to deliver noxious stimuli, regardless of whether it is successful" (1971, p. 10). However the term "attempt" again reinstated the psychological referent for aggression. In one form or another, the psychological instigation to aggression has served as the critical locus of definition in almost all subsequent treatments of the topic (cf. Berkowitz, 1962; Baron, 1977).[3] Even in Zillmann's (1979) extensive (44 pages) and highly behavioristic account of the concept of aggression, it is concluded that the term refers to "Any and every activity by which a person *seeks* to inflict bodily damage or physical pain ..." (p. 33, italics added). "To seek" is, after all, a state of intent, rather than an activity subject to public observation.

While sometimes embarrassing, the fact that aggression as a theoretical construct generally refers to an internal or hypothetical realm has hardly proved lethal to the research endeavor. Often the fact is simply ignored, and investigation proceeds just as if the term refers to a series of observable events in nature. Various behaviors, such as delivering shock to another person or striking a plastic doll, are simply said to be aggressive and no attempt is made to explicate or examine the underlying motivational base of research participants. However, certain investigators have attempted to remain consistent with the definitional presuppositions. In these cases it is generally advanced that the term aggression has *indirect* grounding in observables. As it is said, aggression is a hypothetical con-

---

3 One significant exception is Bandura (1973), who defines aggression as "behavior that results in personal injury and in destruction of property" (p. 4). However, in an attempt to avoid the kinds of attack leveled against Buss' early work, Bandura adds the promise that "social judgment" must determine which acts are to be labeled aggression. This latter proviso creates as many problems as it solves, for any act may be subject to myriad interpretations, and, thus, what counts as "research on aggression" falls victim to whose definitional tastes prevail.

struct and, although such constructs are not subject to direct observation, one may develop reliable indicators of its presence of absence. For example, when an individual verbalizes his motives (e.g., "Yes, I was trying to kill him"), when the action is accompanied by forethought, or when physiological measures indicate a heightened state of arousal, then it may be more justifiably argued that the behavior in question is aggression and not something else (e.g. altruism, religious worship, etc.). In effect, the term aggression does not refer directly to these measures, but its definition is, in part, linked to observables (e.g., aggression is that which is indicated by a statement of intent, physiological arousal, and so on, which statements are then linked to observables).

But let us examine more closely this resort to the argument for indirect, ostensive grounding. Do behavioral scientists of this persuasion have a proper analogy with the physical sciences, such that the term aggression functions much like those of gravity, atoms, and magnetic force? It would not appear so. This contention becomes clear when one attempts to specify the realm of observables to which the indirect indicators refer. For example, what is the ostensive anchoring for a statement of intent? At first glance, it appears that a "statement of intent" refers to just that, an individual's observable utterances. But does it? If such utterances were recorded with a sound spectograph, and the results of this instrumentation compared with various other utterances produced by the individual, would it be possible to differentiate between utterances indicating an intention to aggress, as opposed to other states or intentions? Clearly, the answer is no. A statement of intent to do harm, in terms of its properties as sound, will look like many other sounds emitted by the person. Similarly, if the statement, "I meant to kill him," were analyzed in terms of its phonemic properties, its syllabic order, or its grammaticality, one could scarcely identify whether the statement was an indicator of an aggressive intent or some other. This is to say that it is not the physical properties of the utterance, "I meant to kill him," to which the descriptive phrase "statement of intent" refers. The speaker could issue such words in a shout or a whisper, in script or cursive, in Hebrew or Sanskrit, in proper grammatical form or in slang, and it would still make little difference as to the judgment of intent.

Then to what does this latter range of descriptors, such as statement of intent, refer? If such descriptor are not themselves ostensively grounded, then what are their referents? As it is rapidly surmised, such descriptors bear a strong family resemblance to the term aggression itself. That is, they appear to have ostensive anchors, but this appearance is a misleading product of the tendency to confer ontic status on the world implied by words. When the veil of objectification is rent, we find again that such descriptors refer to other psychological states. That is, one is not, after all, interested in the physical properties of such descriptors, but in the underlying meaning, motive, intention, and the like. One is not concerned with the physical properties of the utterance, "I meant to ...," but with the speaker's *intention* in saying these words – for example, whether he or she was "trying" to give an accurate picture of an intentional state, prevaricating, speaking out of a trancelike stupor, or the like. In the same way, it is not the individual's statements regarding his or her hostility or indicators of physiological arousal that are ultimately significant in determining whether an action was aggressive, but, rather, what these indicators signify about the psychological conditions under which they were produced. A statement of hostility can, after all, be used to mislead, and an accomplished liar can control his or her state of physiological arousal, even when exposed to a lie detector.

One might pursue the possibility of developing measures that could ostensively define measures of these particular states. However, as this procedure is implemented, it is soon realized that one has entered an infinite regress in which every indicator of a psychological state (e. g., meaning, intention, conceptual system, motive, etc.) is itself defined in terms of other psychological states. To measure the meaning behind the statement, "I intended to . . .," would itself require an interpretation of the psychological state necessary to produce the behavioral outcome. In terms of the distinctions developed earlier, this is to say that all statements about aggression, along with a broad range of other person descriptors, are neither directly nor indirectly grounded in observables. Rather, they are linguistically grounded; their definition is exhausted by an inventory of the linguistic contexts in which they are embedded.

The Structural Unpacking of Aggression Discourse

What does the preceding analysis indicate for the study of aggression? First, it seems clear that the concept of aggression should be "deontologized"; that is, the assumption that the term stands in referential relationship to an array of spatiotemporal events must be discarded. In this case one also finds reason to question the function of what has been understood to be empirical research on the genesis of, the conditions giving rise to, the psychological basis for, the physiological inputs to, or behavioral consequences of aggression. Given the lack of worldly events to which such study might be addressed, how is one to consider the outcomes of such work? At the same time, the present analysis indicates the propitiousness of at least two lines of alternative inquiry. The first is closely related to the earlier discussion of the pragmatic aspects of language. As we have seen, language is an important implement for altering or sustaining social pattern; when part of an ongoing discourse, the term aggression can often be a potent instrument of influence or control. Inquiry is invited, then, into the pragmatics of the language of aggression. How is this particular construction of persons to be achieved in the social sphere and with what effects? From whence is its power in social interchange derived? Openings into these issues have been made in other contributions to the present volume. For example, Mummendey, Linneweber, and Löschper (this volume) describe social factors that enter into the negotiation of whether an act is deemed aggressive. And Ferguson and Rule (this volume) and Tedeschi (this volume) explore various normative considerations that may bear on whether an act is labeled aggressive.[4]

---

4 See also Rommetveit's (1979) discussion of "metacontracts" among participants in the determination of meaning, along with Blakar's (1979) inquiry into language as a means to social

The second major thrust of exploration suggested by the present analysis is into what may be termed the grammar of aggression, that is, the rules or conventions governing common discourse about aggression. As we find, aggression is essentially an integer in a language system. Thus, the constraints over what may be said about aggression do not lie in the realm of observation, but in the system of conventions for speaking about it. We cannot easily say about aggression that it goes backward, or travels in circles, is influenced by spirits, or is a form of compulsion. In Austin's (1962a) terms, such utterances violate common "felicity conditions" for speaking about aggression. However, in most social contexts we can say that aggressive tendencies vary in their intensity, that people often aggress because they are frustrated, or that some cultures are more aggressive than others. None of these pronouncements is warranted by virtue of common observation; such warrants are embedded within the common conventions of language use. This is to say that the limits to what science may "discover" about aggression as "a phenomenon" are, for the most part, already lodged within the common conventions of discourse. To elucidate these conventions is to begin to apprehend the limits of what science may generate as "knowledge about aggression." Study of these conventions also serves emancipatory functions, for to gain cognizance over the conventional basis for accepted truth is to remove dependency on such conventions and to invite the creative development of alternatives.

It is to the latter of these programs of inquiry that our major attention will now be given. Later we shall touch on the implications of this analysis for the pragmatics of aggression. The focal attempt at this point is to lay bare certain suppositions embedded within common discourse concerning aggression. This analysis must necessarily proceed with caution, for there are no commonly accepted and well-honed techniques presently available for this purpose. The present analysis does benefit from a variety of preceding inquiries. Within ordinary language philosophy, disquisitions on the language of mind (Ryle, 1949), motivation (Peters, 1958), sense data (Austin, 1962b), emotion (Kenny, 1963), and the like furnish potent demonstrations of the significance of contextual clarification for philosophic inquiry. As Shotter & Burton (1983) have shown, there is also within Heider's (1958) formulation of naive psychology an implicit grammar for action accounting, one that furnishes a rudimentary logic underlying explanations for human behavior more generally. More recently, Smedslund (1978, 1980) has laid out a series of axiomatic definitions of human action from which an indefinite number of theoretical propositions could follow. As Smedslund argues, such propositions are not subject to empirical assessment once the initial definitions are accepted, even when treated as empirical by the scientific establishment. Further, Ossorio (1978) has ambitiously attempted to furnish a set of fundamental dimensions required by linguistic practices in distinguishing human actions from each other. Additionally helpful are the attempts of Ossorio

---

power, and Pearce and Cronen's (1980) analysis of the social management of meaning. The work of Garfinkel (1967) and others within the ethnomethodological tradition (cf. Psathas, 1979) is also germane.

(1981) and Davis and Todd (1982) to develop a *paradigm case* method for determining the set of ordinary language criteria relevant to the use of a given concept.[5]

The method developed for present assessment may be termed *structural unpacking*. It is essentially a formalized rendering of the linguistic conventions necessary or essential for deploying a descriptive term. In this particular case the task is to specify the linguistic conditions commonly pertaining to the deployment of the integer, aggression. It is to convert to a formal structure what is commonly taken to be intelligent discourse about aggression. To speak of the result as a structure is not to link the present analysis with the kind of structuralist inquiries undertaken by Levi-Strauss, Lacan, and others within the French structuralist tradition. In most respects the foundational rationale for the present undertaking is in contention with the universalist leanings inherent in much structuralist writing. Rather, the term structure is used in the present case to call attention to the pervasive character of many current conventions. It is to catch what are essentially evolving patterns of discourse in a single stopframe.

The process of structural unpacking proceeds through the posing of criterial questions for usage of the descriptor in question. Each question attempts to determine a criterion that would enable the common user of the language to distinguish between the appropriateness of using the descriptor in question, as opposed to a range of alternative terms. To consider the structural unpacking of the term apple, a criterial query concerning color would enable most persons to sort between apple and a wide range of competitors. Similarly, the criterial question of "edibleness" would permit one to distinguish between the appropriateness of using the term apple, as opposed to a wide variety of alternatives (e.g., airplanes, hats, stones). As additional criterial queries are posed, the conditions under which it is possible to call an object an apple are progressively narrowed. As criteria of color, shape, size, taste, and so on are added, one gains increasing clarity concerning the linguistic conditions under which it is appropriate to employ the term apple and no other term.

As should be apparent, this form of analysis leaves a certain degree of latitude for the investigator, as there are no definitive rules for what criterial queries may be posed. Thus the investigator is thrust back on his or her own familiarity with the language or must consult other language users. Further, such analyses may be viewed as approaching, rather than reaching, completion over time. There is no obvious means of determining when all possible queries have been established, and, given the continuous evolution of language usage, it may also be supposed that differing criteria may become appropriate over time (e.g., the criterion "city" would presently include "The Big Apple"). Finally, many descriptors are essentially polysemous, that is, they are embedded within several differing structures. For example, the descriptor "cool" is employed in at least two,

---

5 Significant foundational work for the present undertaking is also represented in the contributions of Burke (1945), Mills (1940), and Scott and Lyman (1968). And, of course, Wittgenstein's (1980) inquiries into the philosophy of psychology must be considered seminal to all of the above.

quite distinct language contexts. Criterial queries must be sensitive to such structural differentiation.[6]

With this sketch in mind, we may proceed with the structural unpacking of aggression. Although there are differing structures in which this descriptor is featured, let us presently confine ourselves to the structure that seems implicit in most psychological research, in much policy making, and in many contexts of daily life. As an initial criterion, it seems clear that the term is used as a descriptor primarily when talk is about *animate,* as opposed to *inanimate,* beings. That is, the criterion of animation separates aggression from a huge number of other descriptors. Thus we cannot (except metaphorically) speak of the table or the clock as aggressing; we can speak of aggression in humans, primates, insects, and even in the plant kingdom. Although this criterion may appear of scant significance for an initial cutting device, there are two noteworthy implications. First, implied by the process of structural unpacking is the possibility of ultimately developing a taxonomy of structure. Similar to Levi-Strauss's elucidation of structure through the principle of homology, by separating terms according to criteria of broad consequence, family resemblances among descriptors may become apparent and functional or pragmatic equivalencies clarified. Thus, for example, the term *aggression* shares certain characteristics with terms like *love* and *respect;* this class of descriptors is essentially reserved for animate beings. Further, this initial criterion appropriately recognizes and extends the distinction developed within the philosophy of social science between action and behavior (cf. Taylor, 1964). As argued in this domain, descriptions of human action necessarily make certain presumptions that pertain to human beings, but not to inanimate objects. Those descriptors used in the natural sciences are thus inappropriately and misleadingly extended to human action. Sense cannot be made of human action, as the behaviorists believed, by employing a language reducible to physics. As the present analysis suggests, however, the prerequisite distinction among descriptors should be that of animate vs inanimate, and not human vs nonhuman. Much of the language for person description is equally applicable to other animal beings.

As a second orienting criterion, it seems clear that the term *aggression* falls into a class of terms used in discourse about *interdependent,* as opposed to *independent,* activity. That is, contemporary conventions generally prevent one from employing the term aggression to communicate about the conduct of a sole individual. One simply cannot aggress unless there is a target for his or her actions. In this regard, *aggression* is structurally similar to terms like *dancing, cooperating, helping,* or *ruling.* The descriptor is functionally dissimilar to terms such as *singing, painting,* or *exercising* – each of which may be applied to the activities of a single animate being. As this distinction indicates, the warrantability of using aggression as a descriptor can depend on characterization of at least two actors. Constraints over the use of the term may be traced both to talk about the actor and the recipient.

Given this latter bifurcation, let us first consider the simpler case, that of the

---

6 See Rommetveit (1979) for discussion of the differential contexualizations of single words.

recipient or potential victim. What must be the case in speaking of the victim of the action if the actor is to be termed aggressive? There would appear to be two major requirements. First, the recipient must derive *pain,* as opposed to *pleasure,* from the actor's conduct. If it were said of the recipient that, although he seemed to be in pain, secretly he enjoyed the actor's behavior, it would be inappropriate to speak of the behavior as aggression. (It might, for example, be viewed as sadomasochistic.) Second, the recipient's pain must be understood as *undeserved* (unjust, inappropriate), as opposed to *deserved* (just, appropriate). If the actor's behavior causes pain and the recipient is deserving, we would not meaningfully call the action aggression. Rather, it would be more appropriate to use such terms as punishment, retaliation, self-defense, correctional action, moral training, teaching, or the like. The use of the descriptor *aggression* may require additional characterizations of the recipient, but for present purposes we may consider these criteria as orienting and foundational.

In turning to the actor (the potential aggressor), three essential conditions are apparent. First, it must be said of the act that it was *intentional,* as opposed to *unintentional.* One cannot thus speak of an unintentional aggressive act. If a hunter believes he is shooting a bear and fells a colleague, it would not be said of the act that it was aggressive. However, if it were said of the same action that the hunter intended to slay his fellow, it would be quite appropriate to describe the action as aggression. Second, the aim of the intention must be that of delivering *pain* or harm to another (or others), as opposed to *pleasure.* Thus, if the agent said of her caustic criticism that it was intended to help the recipient correct his miserable ways and once again find happiness, it could not appropriately be said that she aggressed; rather, it might be said that she was teaching a bitter lesson or taking his future in her hands. And finally, it must be said that the actor's conduct is *unjust* or inappropriate, as opposed to *just* or proper. Even though it may be said of a hangman that he intends to deliver harm to another, it would not typically be said that his conduct is aggressive. Rather, he is doing his duty, serving the public, or helping to dispense justice. If the same individual were to hang someone not deemed a villain by society, his actions might be called wantonly aggressive.

Thus far a series of seven criteria have been proposed, each of which appears stipulative with respect to discourse on aggression. These seven criteria have been collected in Figure 4.1. As proposed, virtually all of these criteria must be either implicit or explicit as one speaks of an act as aggressive. Let us term this array of criteria a *structural nucleus.* Its nucleic property derives from the fact that, at least in the ideal case, virtually all the indicated properties, and only these properties, must be assumed. For purposes of defining aggression the set is complete and autonomous within itself. (Further analysis would surely reveal necessary additions to the nucleus, but for purposes of the present analysis the seven criteria will prove sufficient.) It should further be noted that the terms of the nucleus are essentially redundant. That is, once an action has been called aggressive all of the constituent features of the nucleus can be employed felicitously in speaking of the actor and the recipient. Or, once the actor and the recipient have been described in these terms, no new information is added by calling the action aggressive.

Fig. 4.1. Structural nucleus for aggression

Yet, while the structural nucleus lays out the warranting conditions for the term aggression, the schema must be expanded in an essential way. The predicate aggression may figure in many accounts of action in which the definitional equivalents do not occur. It may properly be said, for example, that, "The man was angry and therefore aggressed," while "He was happy and therefore aggressed" borders on cultural nonsense. To use the term "angry" in the preceding context is thus to increase the probability of describing the action as aggressive, while the term "happiness" inhibits the ascription. Yet, neither anger nor happiness is a constituent element of the structural nucleus. Likewise, if it were said that, "Mary criticized her friend Martha," one would be less inclined to view this criticism as aggression than if Martha was said to be Mary's enemy. In effect, the probability of employing the term aggression depends not only on placing it within the context of its definitional nucleus, but on additional attributes of the linguistic context, as well. The task that now confronts us is how to account for these latter effects in terms of the structural analysis proposed here.

To answer this query let us return to examine the relationship between the *primary term* in the nucleus "aggression" and the *secondary terms* on which it relies for its definition. This relationship is of one to many, in the sense that the primary integer requires a set of greater than one in order for parity or identity to be achieved. Thus, the secondary terms possess the meaning "aggression" only when employed as a group. However, this leaves open the definition or meaning of the secondary integers in isolation. As is quickly seen, the meaning of these terms is also subject to structural unpacking. That is, each of the single terms comprising the secondary integers in the preceding analysis may be viewed as a primary predicate within a separate structural nucleus. The "intention," "harm," "just," "animate," and so on stand subject to the same unpacking procedure as did aggression. Each possesses secondary equivalencies that serve to define or to establish the warrant over its usage. The exposition of this relationship between primary and secondary terms, the latter of which stand in a primary to secondary relationship with still further terms, and so on, may be viewed as *vertical unpacking*.

The major integer in question (i.e., aggression) is featured as the vertex of the network and is accompanied by a descending order of related structural nuclei.

Vertical unpacking may be contrasted with the process of *horizontal unpacking*. In this case, attention is drawn to the fact that each of the secondary integers in the initial nucleus may also be featured as a secondary integer in a range of other nuclei. Thus, for example, the term unjust, which partially contributes to what is meant by the term aggression, is also a secondary integer in a variety of other nuclei. If it were said of a person, "He underpaid his workers," the descriptor "underpaid" would be partially defined by the secondary term unjust. Similarly such descriptors as cheating, exploiting, and oppressing would undoubtedly contain the term unjust as a part of the secondary retinue. It follows from this line of reasoning that each structural nucleus possesses conventionally based and overlapping attachments to a range of other nuclei, both in the horizontal and vertical dimension, and these to still others. In effect, each structural nucleus is a constituent of an interrelated (and indefinitely extended) complex.

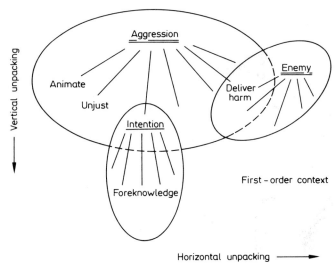

Fig. 4.2. Constituents of the aggression nucleus as embedded in other nuclei

A partially unfolded schematic of the broader linguistic context of the aggression integer is featured in Figure 4.2. For present purposes let us focus attention on the relationship between the aggression nucleus and several structures to which it is immediately related. We may term these immediately attached nuclei, both within the horizontal and vertical planes, nuclei of the *first-order context*, inasmuch as the secondary terms of the structural nucleus in question are either defined by, or are featured within, these nuclei. Now, if the term aggression is made more or less probable, depending on whether secondary integers within its own nucleus are employed in a descriptive account, then the use of these secondary integers may be appropriately governed by the occurrence of the various terms within the first-order context. To illustrate, in the vertical dimension a critical criterion of whether an individual possesses an intention to act is whether he or she has *foreknowledge* of results. If it is said that a person did not know his or her actions would have a particular effect, it could not properly be said that the effect was intentional. Thus, to say that an individual had no foreknowledge indirectly affects the warrantability of calling the person aggressive. To move now in the horizontal direction, it was said that aggression was partially defined by the criterion of delivering harm. Yet, within the first-order context we find the integer "delivering harm" is also featured as a secondary integer of other nuclei. For example, the definition of "enemy" would undoubtedly include, as a definitional criterion, the delivery of harm, and not benefit; in contrast, to be a friend is, by definition, to be someone to whom benefit, rather than harm, is given. Given these linkages, it becomes more appropriate to say that criticism of an enemy is aggressive, while criticism of a friend is not.

As we may conclude, whether or not one is permitted to speak of an individual or group as aggressing depends not on the physical characteristics of the action in question, but on the linguistic context in which the term is embedded. This context is not only composed of the secondary predicates (e. g., intention, delivering harm, unjust) that go into defining aggression by conventional standards, but also by the context in which these predicates are embedded. Thus, when an individual uses terms within the broader context (either horizontal or vertical), such terms influence the likelihood of one's making sense with terms in the primary context. Whether descriptive terms in the context have less warranting power than the constituents of the nucleus itself, or whether those in the vertical dimension have more power than those in the horizontal, remain interesting questions for future inquiry. The major purpose of this treatment has been to furnish an analytic account of the linguistic grounding of person descriptors in general, and aggression in particular.

On the Negotiation of Aggression

A distinction was initially drawn between two ways in which language creates reality, the first by objectification and the second by pragmatic deployment. Although analytically separate, these modes of world construction are closely related. One's ability to achieve social effects depends on the structure of objectifications, and this structure is realized only within the nexus of pragmatic encounter. It is to the former of these dependencies, that of pragmatics on structure, to which we must at last turn, not only because it saves the previous analysis from becoming an arid formalism, but because the major focus of this volume is on aggression as social process. It seems clear from the present analysis that virtually all social patterns stand vulnerable to the ascription of aggression; likewise, virtually any other predicate within the vocabulary of person description can be used to account for what is taking place. As argued, the selected descriptor tells us virtually nothing about the features of the ongoing movements; it in no way reflects the state of nature. Rather, the selection of descriptors may principally be viewed as a performative, or an integer within a flexible though partially structured social sequence. In the fashion of Putnam (1978) or Habermas (1971), one might wish to argue that person description is quintessentially "interest relevant" in this respect. However, to speak of the interests, designs, or intentions underlying the selection of descriptive terms is once again to enter the structural labyrinth.

Rather than being drawn into futile conjecture about purpose, interest, or intent, it seems more fruitful to scan the sorts of rhetorical structures in which the term aggression most frequently figures. That is, we may ask what follows "logically" (in terms of conventional sense-making) if an individual or group is said to be aggressive. What else is one then permitted to say or do?[7] Although there are a number of possibilities, at least one of the most commonly employed sequences is terminated in punishment. As we have seen, by common definition, aggression is an unjust action. And within common language conventions, unjust or unfair acts are deserving of retribution. They may be set aright by punishing the agent. Thus, to attribute aggression commonly furnishes one with the right, if not the duty, to bring harm to the actor. And these reciprocal acts, because they achieve justice, are not easily subject to the ascription of aggression. More generally, then, to call an act aggressive is to generate a right to control or punish the agent and to do so in a way that is in itself not subject to retribution.

Given the attempt to attribute aggression to a person or a group, the outcomes of the structural unpacking process prove highly informative. Such outcomes first inform one of the range of linguistic integers that must be invoked if the ascription is to be justified or sustained. If one is to explain why an action is to be distinguished as aggression, as opposed to a host of competitors, structural unpacking specifies how this must be done. First, it may be anticipated that the adversary will draw from the range of secondary constituents of the definitional nucleus. For example, one attempting to demonstrate how the Russian presence in Afghanistan is an act of aggression might wish to argue that acts of violence were perpetrated against the people (bringing harm), such attacks were planned (intention), the Russians lacked legal or moral right to invade (injustice), and so on. Although it might first appear that these various elaborations furnish a factual warrant for calling the Russians aggressive, a more sensitive analysis reveals that for purposes of ascription, this elaboration adds no new information. That is, once the act has been termed aggressive, all the components hold true by definition. To say that the act is aggressive *because* it brings harm, is intentional, and so on is essentially to say that the act is aggressive because it is aggressive. Or, in terms of literary analysis, statements regarding intention, injustice, and so on serve as tropes, or figurative substitutes, for aggression.

As the earlier analysis also makes clear, the attempt to justify or sustain the ascription of aggression will not be limited to employing terms from the structural nucleus alone. Rather, the adversary may also usefully draw predicates from the first-order context, both horizontal and vertical. For example, one may draw from the vertical context in trying to demonstrate the validity of intention (in

---

7 It is not, thus, as Grice (1957) and Searle (1970) propose, that the fundamental goal of the listener is to determine what the speaker *intended* him or her to understand by a sentence. Rather, as indicated here, the task faced by the listener is to assess the *action consequences* (including linguistic) that would result, should the sentence be accepted.

the primary nucleus) by maintaining that the Russians knew in advance that bloodshed would be required. In this case the adversary would essentially be citing a definitional component (possessing cognizance) that contributes to the definition of what it is to have an intention. Or, to draw from the horizontal context, one might argue that the Russians were not traditional friends of the Afghans or decided unilaterally to invade the country. The former use of the term "not friends" gives conventional warrant to the criterion of "harm-doing" in the primary nucleus; to speak of "invasion" is to increase the warrant for "unjust" in the primary nucleus, as to invade is, within a particular definitional nucleus, an unjust act. Again we see that the resort to terms in the first-order context does not add essential information to the ascription. All such terms are redundant with the ascription of aggression. The justification is essentially a rhetorical one, amounting to a multiplicity of overlapping assertions.

Yet, in cases of ascribing aggression it is often difficult to obtain univocal agreement. Others may be engaged in differing ascriptional projects, most particularly if they are the targets of the initial labeling. As Mummendey and her colleagues (Mummendey, Bornewasser, Löschper, & Linneweber, 1982) have said, "It's always somebody else who is aggressive." And with good reason. As we have seen, the ascription of aggression renders one vulnerable to punishment. At this juncture the results of the structural unpacking process again take on a predictive function. In this case they inform one of the linguistic grounds upon which negotiation is likely to occur. It is useful in this case to view the participants in such negotiations as ontological adversaries, each attempting to vindicate a particular reality while vitiating that of the opponent.[8] Thus, the supporter for the Russian presence in Afghanistan might first assert an alternative definition of the situation (e.g., self-defense, giving support to a teetering government) and then demonstrate the invalidity of the elements of the structural nucleus and its relatives within the first-order context. The attempt might thus be to show how the common Afghan is experiencing no harm, but is being aided, that the governing power has extended an invitation to Russia, and that there was no intention to do harm to the nation – in each case undermining one of the definitional supports for the aggression ascription. Or, a tactical move may be made within the first-order context. It may be maintained (vertical context level) that the warfare has been conducted against rebel forces attempting to overthrow the justly constituted government, and thus the act is not unjust. And, within the horizontal context it might be ventured that Russia has considered itself an ally and protector of its close neighbor, thus undermining the validity of the "harm-doing" constituent of the primary nucleus. In effect, each adversary may be expected to move from one sector of the aggression nucleus to the associated context to sustain a given position regarding the act.

---

8 For a relevant analysis of the negotation of conflicting metaphors, see Lakoff and Johnson (1980).

Summary

This chapter attempts to lay the groundwork for understanding discourse on aggression. As it is shown, the term itself is defined by, or can be equated with, a set of criterial attributes. The term and its attributes are termed a structural nucleus. Further unpacking reveals that each criterial attribute is embedded within still other nuclei. Such terms may figure as predicates to be defined by further attributes within a nucleus or as defining characteristics for still other predicates. It is further proposed that virtually all that may sensibly be said about aggression, whether in the conduct of science or social relationships more generally, can be derived from the full unpacking of the language conventions. As proposed, such language conventions cannot, in principle, be corrected or corroborated by observation of human behavior.

References

Austin, J. L. *How to do things with words.* New York: Oxford University Press, 1962. (a).
Austin, J. L. *Sense and sensibilia.* London: Oxford University Press, 1962. (b).
Bandura, A. *Aggression. A social learning analysis.* Englewood Cliffs, N. J.: Prentice-Hall, 1973.
Bandura, A., & Walters, R. H. *Social learning and personality development.* New York: Holt, Rinehart & Winston, 1963.
Baron, R. A. *Human aggression.* New York: Plenum, 1977.
Berkowitz, L. *Aggression: A social psychological analysis.* New York: McGraw-Hill, 1962.
Blakar, R. M. Language as a means of social power. In R. Rommeteveit & R. M. Blakar (Eds.), *Studies of language, thought and verbal communication.* London: Academic Press, 1979.
Burke, K. *A grammar of motives.* New York: Prentice Hall, 1945.
Buss, A. H. *The psychology of aggression.* New York: Wiley, 1961.
Buss, A. H. Aggression pays. In J. L. Singer (Ed.), *The control of aggression and violence: Cognitive and physiological factors.* New York: Academic Press, 1971.
Davis, K. E., & Todd, M. J. Friendship and love relationships. In K. E. Davis & T. Mitchell (Eds.), *Advances in descriptive psychology* (Vol. 2). Greenwich, England: JAI Press Inc., 1982.
Dollard, J., Doob, L. W., Miller, N. E., Mowrer, O. H., & Sears, R. R. *Frustration and aggression.* New Haven, Conn.: Yale University Press, 1939.
Feshbach, S. The function of aggression and the regulation of aggressive drive. *Psychological Review,* 1964, *71,* 257–272.
Garfinkel, H. *Studies in ethnomethodology.* Englewood Cliffs (NJ): Prentice Hall, 1967.
Gergen, K. J. *Toward transformation in social knowledge.* New York: Springer, 1982.
Grice, H. P. Meaning. *Philosophical Review,* 1957, *64,* 377–388.
Habermas, J. *Knowledge and human interest.* Boston, Mass.: Beacon Press, 1971.
Heider, F. *The psychology of interpersonal relations.* New York: Wiley, 1958.
Kaufmann, H. *Aggression and altruism: A psychological analysis.* New York: Holt, Rinehart & Winston, 1970.
Kenny, A. *Action, emotion & will.* London: Routledge & Kegan Paul, 1963.
Lakoff, G., & Johnson, M. *Metaphors we live by.* Chicago: University of Chicago Press, 1980.
Lubek, I. A brief social psychological analysis of research on aggression in social psychology. In A. R. Buss (Ed.), *Psychology in social context.* New York: Irvington, 1979.
Mills, C. W. Situated actions and vocabularies of motive. *American Sociological Review.* 1940, *5,* 904–913.
Mummendey, A., Bornewasser, M., Löschper, G., & Linneweber, V. It is always somebody else who is aggressive. A plea for a social psychological perspective in aggression research. *Zeitschrift für Sozialpsychologie,* 1982, *13,* 177–193.

Ossorio, P. G. *"What actually happens."* Columbia, S. C.: University of South Carolina Press, 1978.

Ossorio, P. G. Outline of descriptive psychology for personality theory and clinical applications. In K. E. Davis (Ed.), *Advances in descriptive psychology* (Vol. 1). Greenwich, Conn.: JAI Press, Inc. 1981.

Pearce, W. B., & Cronen, V. E. *Communication, action and meaning: The creation of social realities.* New York: Praeger, 1980.

Peters, R. S. *The concept of motivation.* London: Routledge & Kegan Paul, 1958.

Psathas, G. (Ed.) *Everyday language: studies in ethnomethodology.* New York: Irvington, 1979.

Putnam, H. *Meaning and the moral sciences.* London: Routledge & Kegan Paul, 1978.

Quine, W. V. O. *Word and object.* Cambridge, Mass.: M. I. T. Press, 1960.

Rommetveit, R. *Words, meanings and messages.* New York: Academic Press, 1968.

Rommetveit, R. Deep structure of sentences versus message structure. In R. Rommetveit and R. M. Blakar (Eds.), *Studies of language, thought and verbal communication.* London: Academic Press, 1979.

Ryle, G. *The concept of mind.* London: Hutchinson, 1949.

Scott, M. B., & Lyman, S. Accounts. *American Sociological Review,* 1968, *33,* 46–62.

Searle, J. R. *Speech acts: An essay in the philosophy of language.* London: Cambridge University Press, 1970.

Shotter, J., & Burton, M. Common sense accounts of human action: The descriptive formulations of Heider, Smedslund, & Ossorio. In L. Wheeler (Ed.), *Review of personality and social psychology* (Vol. 4). Beverly Hills, Calif.: Sage, 1983.

Smedslund, J. Bandura's theory of self-efficacy: A set of commonsense theorems. *Scandinavian Journal of Psychology,* 1978, *19,* 1–14.

Smedslund, J. Analysing the primary code. In D. Olson (Ed.), *The social foundations of language: Essays in honour of J. S. Bruner.* New York: Norton, 1980.

Taylor, C. *The explanation of behavior.* London: Routledge & Kegan Paul, 1964.

Wittgenstein, L. *Remarks on the philosophy of psychology* (Vols. 1 & 2). Oxford: Basil Blackwell, 1980.

Zillmann, D. *Hostility and aggression.* Hillsdale, N. J.: Lawrence Erlbaum, 1979.

Chapter 5

Aggression: From Act to Interaction

Amélie Mummendey, Volker Linneweber, and Gabi Löschper

The motivation for psychologists to do research on aggression seems to be rather simple and homogeneous: nearly every introduction or preface of a monograph on aggressive behavior refers to the social importance and urgency of providing knowledge on the origins of this kind of behavior and contributing to the solution of social problems concerning aggression and violence.

Aggression as a research topic has found its way into psychology not as a scientifically generated question, but as a problem directly taken over from everyday experiences. This kind of origin brings with it certain difficulties in transforming the everyday problem into a concept which is suited for the theory and methods of psychology. Because of these difficulties, theoretical and empirical work on aggression has always been accompanied by a discussion about the adequacy of definitions and operationalizations of aggressive behavior at the time. The familiar discussion about intent as a necessary definition criterion may be considered the result of the perceived discrepancy between aggression as conceived in everyday life and conceptualizations about aggression in psychology (Buss, 1961; Kaufmann, 1970; Werbik, 1971; Zillmann, 1979). This discrepancy only becomes obvious if there is a need to recur from time to time to real-life events in question.

People who are worried about aggression and interested in its origins have probably in mind events like robbery, homicide, genocide, and war. Psychologists try to overcome the difficulties of instigating such kinds of behavior in the laboratory or field by creating particular experimental paradigms which reduce these events to a certain individual reaction in a special social situation. This reaction usually has the following characteristics: (1) it produces aversive consequences for the recipient, or better yet, the experimenter thinks that the subject is made to believe that these consequences are caused by this kind of reaction (e. g., applications of electric shocks, noise, etc.) and (2) the subject has the possibility of choosing between the admission of more or less intense aversive stimuli or of a smaller or larger number of such stimuli. This possibility of choice is the basis for the experimenter to infer the subject's intent to apply aversive stimuli. According to the consensual definition of aggression as a person's action performed with intent to hurt or harm another person (Kaufmann, 1965; Berkowitz & Donnerstein, 1982), this kind of operationalization of aggressive behavior seems to be adequate. Until the present, most of experimental work on aggression was done using this experimental paradigm (e.g., Berkowitz, Cochran, & Embree, 1981).

This is not the place to return to the discussion about the usefulness of the "Buss-Berkowitz aggression paradigm" (cf. Tedeschi, Smith, & Brown, 1974; Hilke, 1977; Berkowitz & Donnerstein, 1982). The central question in this debate is whether subjects "really" behave aggressively when they participate in these kinds of experiments. Because of the special relationship between real life and scientific concepts of aggression, it seems to be necessary to come back to the question of what aggression "really" is, or better, which crucial points are meant when people think of problems with aggression. After some discussion about the characteristics of what is normally meant by "aggression," we shall propose some ideas on how to approach the phenomenon of aggression in psychology without losing its most socially problematic, and therefore most essential and interesting, characteristics. As will be shown, it is not a question of whether or not to perform laboratory experiments on aggression; the question is rather what kind of theoretical conceptualization should be set up for performing the experiments within a modified theoretical framework.

Looking at the immense variety of aggressive events there can be extracted the following pattern common to all these phenomena. When we identify an event as aggression we perceive the following facets.

1. A person (the recipient or victim) runs into conditions which (s)he wants to avoid. This means the recipient suffers harm or injury or would have suffered it if chance or luck wouldn't have prevented it.
2. Under ordinary circumstances, these conditions would have been avoidable. They wouldn't have occurred if another person or agent (the actor or aggressor) hadn't created them (cf. Arendt, 1970 and Bernstein, 1980 for more details about the crucial role of avoidability of aversive events in violence and hostility). The actor thus decided to perform an action causing these conditions, albeit having further alternatives of actions which might have produced less aversive conditions for the recipient. This means *the actor caused the aversive consequences and is responsible for them.*
3. To have created these aversive conditions for this recipient within that particular situation is evaluated as violating situationally relevant norms. Norm violation is apparent, at least from the victim's perspective, and perhaps from the perspective of an outside observer as well. In the eye of the victim, her/his aversive state was caused by the aggressor, although (s)he could, and therefore should, have avoided it.

It is postulated that any event missing one or more of the above-mentioned facets wouldn't lead the observer to identify an aggression. The term "observer" is meant to include every person evaluating an event (except to actor). That does not mean that any observer would check thoroughly to verify the presence of all facets before evaluating an event as aggressive. It is assumed, moreover, that –

analogous to the tendency for completion towards the *gute Gestalt* – the observer will rely on empirically as well as ideologically derived inferences about one or the other facet.

Similar to the above-mentioned facets, which in combination form the picture of an aggressive event, other authors have proposed harm, intent, and norm violation as criteria for the definition of aggression (Tedeschi & Lindskold, 1976; Rule & Nesdale, 1976; Da Gloria & De Ridder, 1977; Werbik & Munzert, 1978). This resulted in the conclusion that the concept "aggression" is not a descriptive but an evaluative one, which leads to the conclusion as Herrmann (in press) put it: "Bestimmt man aggressives Handeln u.a. mit Hilfe vorliegender Normen bzw. der Abweichung von diesen Normen, so ist beobachtbares Verhalten gleicher Art, entsprechend der Änderung sozialer Normen, allenfalls einmal als aggressiv und dann wieder als nicht-aggressiv zu klassifizieren: Eine invariante Bestimmungsregel für aggressives Handeln kann somit zur variablen Bestimmung einer konstanten Verhaltensklasse als aggressiv vs. nicht-aggressiv führen, insofern sich die Normen entsprechend ändern." (p. 18)[1]

Consistent with this orientation, three different approaches for aggression research can be discerned. The first is represented by the action-theory approach proposed by Werbik, which focuses on the aggressor and her/his action. Opposite to this is the approach via coercive power proposed by Tedeschi and co-workers who see aggression as a label resulting from perceptions by the victim. Finally, differing from these two concepts, is the concept of aggression as social interaction proposed by the present authors, which aims at including both actor and recipient.

According to Werbik (1971, Werbik & Munzert, 1978), when studying aggression one must be absolutely certain that the individual prepares and performs an action which fulfils the criteria. The researcher must be certain that the subject plans and performs an action that carries the subjective meaning *(Sinn)* of producing harmful or destructive effects for another person. To preserve the concept of aggression as a specific form of individual behavior, Werbik tries to make sure that the necessary aspects, like harmful intent and norm violation, are present quasi-within the individual and that the individual really acts aggressively. To establish this certainty for the experimenter, Werbik proposes certain methods of standardized action descriptions provided by the subjects themselves, which will not be described here (cf. Werbik, 1971 for more detail). What should be stressed in this context is the following. Although the evaluative character of the category "aggression" is emphasized, this approach restricts itself to the individual level of analysis. By something like a trick, all the information necessary for identifying an action as aggressive is stuffed into a model of individual action. The individual produces, by a combination of behavioral and verbal performances, the

---

1 "If aggressive actions are defined with respect to relevant norms or the violation of these norms, observable behavior of the same kind will be classified according to the change of these norms either as aggressive or nonaggressive: An invariant definitionrule for aggressive actions may lead to a variable classification of a definite category of behavior either as aggressive or nonaggressive, if the relevant norms will change."

critical action. These performances are treated then like usual objective data for description and analysis.

Tedeschi and co-workers propose the opposite approach. Starting from the conclusion that aggression is not a purely descriptive but an evaluative concept, they try to purge the description of evaluative ingredients. For the objective description of behavior, a classification system in terms of different kinds of coercive power is proposed. The term aggression is reserved for describing results of judgmental processes by the observer, thus reducing the whole thing to a question within the area of social or person perception. Thus the classical field of aggression research splits into two areas: the study of conditions of coercive influence on individual behavior, which is described by objective observation, and the study of conditions for judging and evaluating different kinds of coercive behavior as aggressive or not.

The conceptualization of aggression proposed here chooses a third way. It is neither possible nor necessary for the researcher to make sure that the acting individual perceives the critical criteria "intent to harm," "aversive consequences", and "norm violation" as characteristics of his/her own action. The observer evaluating an action as aggressive uses different information which transcends that provided only by the actor. There is a lot of literature about the conditions influencing the inference of intent of an action (cf. Pettit, 1978, 1981; Harris & Harvey, 1981; Lalljee, 1981). Besides that, the observer who infers intent from an actor may be completely sure about the appropriateness of this inference, although objectively it is impossible to know whether this intent actually *existed within the actor.*

To be able to make inferences about the intent of an action, the amount of harm resulting from this action, and the violation of a norm, the judge needs information provided by the agent and the victim of the action, and additionally the situational context within which the action takes place. The situational context provides hints for the evaluation of the appropriateness of the critical action. Information about the interaction process, i.e., events which preceded the critical action and prognoses about what will happen as a consequence of this action, are taken into consideration as well.

But why then not follow the way proposed by Tedeschi and co-workers? To explain the arguments for our third way it is useful to go back to the more general problem, "aggression" perceived beyond the conceptualizations of psychologists, up to now. If we think of problematic cases of aggression we rather seldom look upon a single act but, rather, upon a sequence of hostile and dangerous exchanges of mutually hurtful actions. And even if we single out one spectacular action, we at least implicitly examine the events preceding the event in question. When we talk about an especially disgusting and brutal action, we usually miss a preceding event which could have given sense to the occurrence of the critical action in terms of revenge for provocation or something similar. This means that if we deal with aggression, we deal with a certain kind of social interaction between at least two individuals (or two parties) in a specific social situation.

What is meant by interaction is, as Graumann (1972) wrote: "... weniger neu entdeckte Phänomene als eine neue Sichtweise ... Phänomene und Sachverhalte, der Sozialpsychologie des längeren als wissenschaftliche Probleme bekannt, werden in ihrem "interaktionalen" und "kommunikativen" Charakter erkannt und neu analysiert. . . . (Es hat) den Anschein, daß die Verwendung der Begriffe Interaktion und Kommunikation der theoretisch gefährlichen Polarisierung Individuum vs. Gesellschaft entgegenwirkt, indem sie stärker, als es früher der Fall war, das, was *zwischen* Mir und dem oder den Anderen, zwischen Individuum und Gruppe oder Gesellschaft geschieht, zu artikulieren gestattet." (Graumann, 1972, p. 1110)[2] This is exactly what needs to be conveyed by the conceptualization of aggressive behavior.

"We shall use the term "interaction" to refer to any set of *observable* behaviors on the part of two or more individuals when there is reason to assume that in some part those persons are responding to each other . . . What all these observable forms of interaction have in common is a *sequence of behaviors* on the part of two or more persons." (Newcomb, Turner, & Converse, 1965, p. 3)

When we apply these more general descriptions to the phenomenon "aggression," a first summary of its central facets results. There is a sequence of observable behaviors performed by two or more persons. These behaviors refer to each other. To describe and analyze such kinds of interactions, a mere consideration of the contingency of two actions wouldn't be sufficient. Although, of course, it would be possible to isolate the *individual contribution* to such an interaction and to look at its particular contingencies (this is exactly what classical aggression research does), this procedure misses the essential characteristics of the phenomenon "aggression." As Graumann (1979) put it for interaction in general: ". . . Das untersuchte interaktionale Phänomen ist . . . zwar nicht "mehr" aber etwas anderes als die Summe dieser Bedingungen oder "Anteile"." (p. 294)[3]

Like the generation of a chemical compound, aggression is seen within the present approach, as a "new" phenomenon generated from a particular combination of necessary elements: two persons or social units, one person performing an action, another person realizing aversive stimulation, locating the cause of this stimulation in the first person; the first person perceiving good reasons for performing this action, from his/her point of view the choice of this particular action being optimum compared with his/her temporal and local circumstances; the net balance showing this choice as *personally* appropriate according to the definition of the situation. Opposite to the first person or actor, the second person or recipient imputes to the actor that (s)he had or should have had fewer aversive alternatives at her/his disposal in this situation, so that exactly the action that was performed would have been avoidable.

---

2 "What is meant by interaction is . . . not so much the discovery of new phenomena but a characterization of a new perspective . . . Phenomena and circumstances . . . are realized with regard to their "interactional" . . . character, and are to be analyzed in a new way: [It seems] that the application of interaction and communication concepts works against the theoretically dangerous polarization of individual versus society by providing the possibility – now more than it has been the case before – to articulate what is happening between Me and the Other(s), between individual and group or society".

3 "The interactional phenomenon looked upon is . . . not "more" but something else than the sum of these conditions or "parts"."

Thus, the two participants, actor and recipient, disagree in their judgments of situational normative appropriateness of the critical action. The actor evaluates it, compared with the particular circumstances, as more appropriate than the recipient does. This includes the proposition that, as Tedeschi and co-workers already put it, the identification of an action as aggression results from a special interpretation and evaluation of an action within a certain context. The present approach goes one step further. The interpretation and evaluation is made more precise by relating it to the judge's *perspective,* which is specific to the *typical positions within the present kind of interaction, i. e., actor or aggressor and recipient or victim.* Thus, the often-uttered critique of conceptualizations of aggression that are based on social judgment, i. e., that there are "distorting factors" (e. g., ethical considerations) involved which "constitute unstable reference points" (Zillmann, 1979, p.38) wouldn't be correct in the present conceptualization. As one of the present authors put it elsewhere:

"Dazu muß die dafür genuine Divergenz der Bezugspunkte der Beteiligten berücksichtigt, nicht durch Ausblenden des einen Teils negiert werden: In diesem Sinne ist der Bezugspunkt in einem interaktiven Konzept gerade nicht instabil (im Sinne von nicht vergleichbar) . . . Der Bezugspunkt ist in einem interaktiven Konzept in definierter d. h. festgelegter Weise, abhängig vom sozialen Kontext der Interaktion und der Position der an dieser Interaktion Beteiligten, variabel." (Mummendey, 1982, p.332)[4]

To come back to the necessary ingredients of the product "aggression": The described units of interpersonal interaction are linked together in a temporally extended sequence. The interpretation of a specific action as aggressive has an impact upon the particular development of the interaction process. If the behavior of the actor is identified as aggressive, then the recipient, perhaps corresponding to the norm of negative reciprocity, considers it his/her right to respond to this aggressive assault in an "aggressive" manner, "to pay back the same with the same" (Lagerspetz & Westman, 1980). Unfortunately, there is not necessarily agreement about what is "the same" between the participants involved.

The form of progress of an aggressive interaction is crucially influenced by the mutual interpretations and judgments of the behavior in question. These interpretations give hints and reasons for the participant's choice of reciprocal actions. If there is interest in the analysis of the different forms of interaction process – whether a continuation or even an escalation will occur, will the process terminates, will there be a compensation of the victim's injury by the actor – the primary question is about the qualities of the formulated definitions of the situations and the judgment of behavior patterns as "aggressive".

It is assumed that divergencies in the interpretation of the different actions

---

4 "... The genuine divergence of point of reference of the participants must be taken into account and not denied by omitting one part of them: In this sense of meaning, the point of reference within an interactive conceptualization is not instable (in the sense of not comparable). ... The point of reference within an interactive conceptualization is in a definite and systematic way variable depending upon the social context and the position of the participants of the particular interaction."

within the sequence to be described will not occur only because of disagreements between the participants or observers when looking upon an identical definite event; there will also be differences in the segmentation of the sequence into discriminable episodes. Thus, the participants, changing their positions during the course of interaction, i. e., being actor and victim respectively, will differ in the perception of the boundaries of the episodes, especially in the identification of the *beginning* of an episode or in defining what is initial event and what is reaction. The relativity of judgments and interpretations includes, besides the evaluation of an incident, the structuring and segmentation of the sequence, i. e., the *construction of events* to be judged (cf. Newtson, 1976; Forgas, 1982).

Here we can again make clear the reasons for not following Tedeschi's propositions to separate the behavioral from the judgmental analysis. By such a separation, the perspective of concepts for the *link between units* of an interaction sequence is blocked up. As we see it, this link is provided by the reciprocal action interpretations attached to the respective behaviors performed. Thus it seems to make some sense to look for a concept which *integrates* the behavioral and the perceptual/judgmental parts.

To sum up the argumentation above, aggression is specified as confrontation between two persons (or social units). The confrontation has an extension in time; there is a sequence of events which the participants structure by defining discriminate segments or episodes. Such episodes consist of an action performed by an actor and aversive consequences felt by the recipient; in each interactional episode, the positions of actor and recipient are provided. The relation between actor and recipient can be characterized as conflict: both actor and recipient have incompatible interests with respect to at least the critical action.

The present concept of aggression as a specific form of social interaction suggests four fundamental aspects: *mutual interpretation, situation context, divergence of perspectives depending on specific positions* (i. e., recipient, actor, or outside observer), and *temporal progress*. In the following sections, four examples of empirical studies concerning details of these four aspects are presented. These studies, each of which is a part of a larger research project, are confined to a limited social field, i. e., the school. The central propositions of the underlying approach include the necessity of studying the influence of consensually valid information about which kind of behavior is appropriate and which is not in what type of situation occurring in the environment of the subjects. For the present purposes, it is far more economic to use a definite and well-describable field which is already established than to simulate a new one. The first study to be reported is concerned with consensual regularities in evaluating an event as aggressive and sanctionable, as a result of different combinations of the definition criteria "intent", "injury", and "norm violation". In this first step, only the judgments made from the observer's perspective are considered. The second and the third studies try to throw some light on the modifying influence of characteristics of the surrounding context and the particular interaction on the judgments in question. The fourth study considers social-consensual conceptions about the progress of aggressive interactions.

Mutually Interpreting Behavior in Aggressive Interactions

The important criteria conducive to the interpretation of critical acts are known as the perceived negative *norm deviation,* the supposed actor's *intent* to harm the victim, and the perceived factual *injury* (cf. Tedeschi et al., 1974; Tedeschi & Lindskold, 1976). Our own studies designed to analyze criteria or conditions of interpreting critical events as aggressive and the possible selection of reactions will be reported here only roughly (for greater detail cf. Löschper, 1981; Löschper, Mummendey, Linneweber, & Bornewasser, 1984). The main interest of this study doesn't concern the conditions under which a negative norm deviation, an intent to harm, and the actual harmful consequences as definition criteria are considered as met in social interactions. Instead of analyzing antecedents of interpretations and attributions, consequences of this process are studied, i. e., how already existing definition criteria influence evaluations of behavior as aggressive and the possible reactions to such behavior.

While the influence of each single criterion – injury (Shaw & Reitan, 1969; Nesdale, Rule, & McAra, 1975), intent (Epstein & Taylor, 1967; Nickel, 1974); and norm deviation (Da Gloria & De Ridder, 1977; Ferguson & Rule, 1983; Stapleton, Joseph, & Tedeschi, 1978) – is well proved, only little is known about their combined effects. However, in several investigations, e. g., concerning moral judgments and conditions of aggressive behavior, it seems that often more than one variable is taken into account – though not in an explicit or systematic way – and confounded with other factors. The fact that often one variable is used as a functional equivalent for another in the above-mentioned investigations, e. g., negative norm deviation and intent to harm, leads to the question of whether some of the criteria or variables are interchangeable or even identical in their subjective meaning. The question arises whether there are systematic patterns in their understanding, as it seems from, e. g., scientific conceptualizations and operationalizations.

Thus two interrelated questions and studies will be reported here: (1) the influence of the definition criteria – norm deviation, intent, and injury – on judgments of critical behavior as aggressive and sanctionable, and (2) the subjective understanding or representation of combinations of these definition criteria.

Norm Deviation, Intent, and Injury as Definition Criteria of Aggressive Behavior

Many experiments have varied the respective factors but only in an unsystematic fashion, thereby confounding them. Nevertheless the specific results provide indirect indications of the possible interaction of norm deviation, intent, and injury and render possible the derivation and formulation of specific assumptions and hypotheses.

Besides the expected main effects of each single variable, it is assumed that both the interpretation of behavior as inappropriate or a negative norm deviation *and* the perception of the actor's intent to injure the victim are necessary conditions for evaluating the critical behavior as aggressive and for hostile responses to it. Behavior that evidently deviates from valid norms and is, in addition, carried out with a clear intent to harm is more likely to be defined as aggressive and to be punished than events that are not unequivocally inappropriate, happen by accident, or fail to meet both criteria. Clearly inappropriate and intentionally harmful acts represent necessary *and* sufficient conditions for evaluating them as aggressive and sanctionable, irrespective of the amount of the factual harmful consequences.

From work on the relation between frustration and aggression, it is known that hostile responses are due more to the lack of justification than to the factual injury or provocation, and more to the intent to harm or frustrate the recipient than to the actual frustration or harm (cf. Berkowitz, 1982; Da Gloria, this volume). Evidently intended and evidently inappropriate acts are therefore respectively assumed to be judged as aggressive and sanctionable, irrespective of the amount of actual injury. If, however, an actor's intent is not obvious or the antinormative character of the act is indefinite, then information about the actual harm serves as an additional indication for the evaluation of the critical behavior, i.e., an increasing amount of injury results in an increasing amount of willingness to punish the critical behavior and to define it as aggressive.

These assumptions were tested using verbal descriptions of aggressive episodes as judgment material. These episodes were constructed on the basis of

Table 5.1. Mean ratings of behavior as aggressive and sanctionable in conditions norm deviation, intent, and injury

		Aggressive		Sanctionable	
		N+	N−	N+	N−
INT+	INJ+	5.12[a]	3.82[c]	4.90[a]	3.33[b]
	INJ−	5.24[a]	3.20[bc]	4.58[a]	1.80[cd]
INT−	INJ+	3.66[bc]	1.54[d]	3.04[b]	1.42[d]
	INJ−	3.00[b]	2.08[d]	2.30[c]	1.84[cd]

ANOVA: Ratings as aggressive
N	$F = 70.13$; $df = 1/390$; $p < 0.0005$
INT	$F = 86.44$; $df = 1/390$; $p < 0.0005$
N × INT × INJ	$F = 6.42$; $df = 1/390$; $p < 0.05$

ANOVA: Ratings as sanctionable
N	$F = 111.65$; $df = 1/390$; $p < 0.0005$
INT	$F = 97.04$; $df = 1/390$; $p < 0.0005$
INJ	$F = 12.55$; $df = 1/390$; $p < 0.0005$
N × INT	$F = 14.07$; $df = 1/390$; $p < 0.0005$
INT × INJ	$F = 6.27$; $df = 1/390$; $p < 0.05$
N × INT × INJ	$F = 15.14$; $df = 1/390$; $p < 0.0005$

Note. Results of $2 \times 2 \times 2$ ANOVA for effects of norm deviation (N), intent (INT), and injury (INJ) on ratings as aggressive and sanctionable. +, high; −, low.
Ratings were done on 7-point bipolar scales labeled aggressive = 7, nonaggressive = 1; sanctionable = 7, nonsanctionable = 1. Cells having different superscripts are significantly different at less than 0.05 level of confidence according to Scheffé test. Superscripts refer to the respective dependent variables separately. Data from Löschper, Mummendey, Linneweber & Bornewasser (1984).

verbal reports of typical aggressive incidents in schools by systematically varying and combining statements of norm deviation, intent, and injury. About 900 pupils of average age 14.6 years each responded to eight episodes representing the eight experimental conditions.

The *manipulation checks* revealed that the manipulation of each variable (norm deviation, intent, and injury) was effective for only one form of two nearly identical series of eight episodes. Beside the expected main effects of each factor – high norm-deviate behavior is rated as more wrong and inappropriate than low norm-deviate behavior, and so on – there emerged additional, unexpected main and interaction effects. These are interpreted as indications for the variation of one variable influencing the perception and subjective understanding of the two other factors – a further hint to analyze the subjective representation of combinations of the three definition criteria.

Therefore the only episodes or experimental conditions to be considered were those interpreted by the subjects in the expected way. This could be seen from the subjects' judgments on rating scales used for the manipulation checks of each variable.

The results of the analysis of variance are presented in Table 5.1.

The evaluation of critical behavior as deserving punishment seems to be far more sensitive to the variation of the factors than the definition of the acts as aggressive. While clearly inappropriate acts and evidently intended ones were more likely to be defined as aggressive and to be rated as sanctionable than only slightly norm-deviate ones or more unintentional ones, an increase in injury only influences the judge's willingness to punish the actor.

The Combination of Norm Deviation and Intent as Sufficient Definition Criteria

Norm deviation and intent strongly determine the evaluations. If an actor clearly violates accepted norms and this is carried out with an intent to harm, the critical behavior is judged as deserving sanctions (cf. Figure 5.1 depicting the interaction of norm deviation and intent). If one or both of the two definition criteria are not met (or only met to a low degree as manipulated here), the judge's willingness to punish the actor is much less dramatic. If neither the inappropriateness of the act nor the actor's intent are stated, sanctions are held to be rather unsuitable. While for both intended and unintended events, punishment is more likely with an increasing amount of norm deviation, the difference is most evident for actions intended to harm.

The negative norm deviation or inappropriateness of an action and the supposed actor's intent to harm are similarly essential for the interpretation of critical behavior. This can be concluded from the graduation of judgments (cf. Figure 5.1).

If both criteria are met, a definite evaluation as sanctionable results. If only one of the two criteria is met, a mean judgment emerges. In the case of neither criteria being met, a low rating is given.

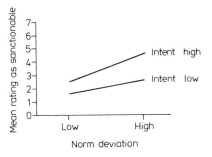

Fig. 5.1. Mean ratings of behavior as sanctionable in conditions norm deviation (high and low) and intent (high and low). Data from Löschper et al. (1984)

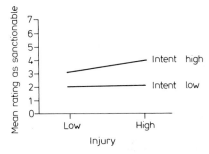

Fig. 5.2. Mean ratings of behavior as sanctionable in conditions intent (high and low) and injury (high and low). Data from Löschper et al. (1984)

In *combination,* norm deviation and intent are necessary and sufficient conditions for the definition of critical acts. There is no difference in defining the behavior as aggressive and rating it as sanctionable between serious consequences and relatively low injury for a victim if *both* criteria – inappropriateness and harmful intent – are met; clearly inappropriate behavior which is carried out with an evident intent to harm is judged irrespective of the amount of factual consequences (cf. Figures 5.3 and 5.4, depicting the higher order interaction of the three factors).

The importance of intent for the interpretation of critical behavior is evident from the interaction of intent and injury (cf. Figure 5.2).

Clearly intended behavior is rated as deserving more punishment with increasing amounts of harm. There is no difference between high and low injury for unintended events.

Apparently, without intent a combination of the definition criteria norm deviation and injury is not sufficient to influence the interpretation of critical behavior. The expected interaction of these two variables was not significant (cf. Table 5.1). To come to a sure definition of an action and a selection of possible reactions, information about the actor's intent must be available.

The Role of Factual Harmful Consequences

From the higher order interaction of all three factors it can be concluded that the described interaction of norm deviation and intent is not independent of the levels of injury (cf. Figures 5.3 and 5.4).

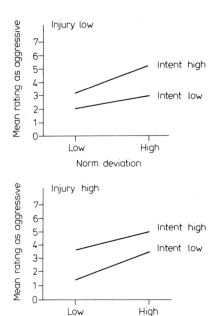

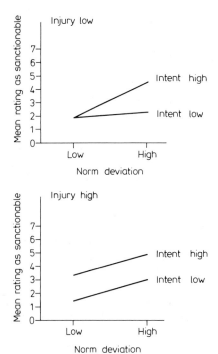

Fig. 5.3. Mean ratings of behavior as aggressive in conditions norm deviation (high and low), intent (high and low), and injury (high and low). Data from Löschper et al. (1984)

Fig. 5.4. Mean ratings of behavior as sanctionable in conditions norm deviation (high and low), intent (high and low), and injury (high and low). Data from Löschper et al. (1984)

If there is only *minor harm* to the victim, both criteria, norm deviation and intent, have to be met to define the critical act as aggressive and sanctionable. With only small indications for the actor's intent, the norm deviation, or both, the critical behavior is likely to be defined as nonaggressive and not to be punished. However, if *serious consequences* result for the recipient, only one of the criteria, norm deviation or intent, evidently suffices for judging the critical act as aggressive and sanctionable.

Thus one can conclude that with a low amount of actual injury the negative norm deviation or inappropriateness of a behavior and the actor's intent to harm are *both* necessary conditions for a pronounced definition of the behavior and the selection of reactions. If the victim is seriously injured, only *one* of the two definition criteria, norm deviation or intent, is necessary for the evaluation and punishment of the critical behavior.

Indications as to form and amount of harm can provide additional information, if an intended action only slightly or not evidently violates accepted norms. With an increasing amount of harm, this behavior is more likely to be defined as aggressive and sanctionable. Information about actual injury, however, *cannot compensate* for either missing information about the actor's intent (concerning the interaction of intent and injury, only intended behavior was influenced by the amount of harm) or missing indications as to the inappropriateness of a behavior, if nothing is known about the actor's intent (the interaction of norm deviation and injury was not significant).

The Subjective Representation of Norm Deviation, Intent, and Injury

The second part of the study concerns the question of whether there are specially frequent configurations in the subjective interpretation of combinations of the three criteria at their different levels. This could be concluded from scientific conceptualizations of the respective variables as well as from results of the manipulation checks. Here the variation of one factor obviously influenced the subjective representation of the other variables of interest. The question arises as to whether there are certain regularities in the subjective meaning and understanding of the combination of the three criteria of norm deviation, intent, and injury.

To analyze this, a configurational frequency analysis (cf. Krauth & Lienert, 1973; Lienert & Krauth, 1975) was performed, which led to the definition of remarkably rare or frequent types. The "objective" variables norm deviation, intent, and injury and their "subjective representation" (determined in the way described before) in the levels "high" and "low" respectively were used in this analysis.

Five configurations show a remarkable discrepancy between the expected and the actual frequencies. Three configurations consist of an identical subjective and objective pattern (cf. Table 5.2); in two, all three variables are on the same level, and in the third norm deviation and intent are on the same level, only injury being at a different level.

Configurations in which the subjective representation is identical with the ob-

Table 5.2. Significantly frequent configurations with the variables norm deviation, intent, and injury in levels high and low for objective condition and subjective representation

			Objective condition			Subjective representation		
Number of cases:	6636							
Number of variables:	6							
Alpha risk:	0.005							
Number of cells:	64		N	INT	INJ	N	INT	INJ
$F_{expected}$:	65.2	$p \leqslant 0.003$	1	1	1	1	1	1
$F_{observed}$:	246							
$F_{expected}$:	65.1	$p \leqslant 0.002$	2	1	1	1	1	1
$F_{observed}$:	248							
$F_{expected}$:	75.9	$p \leqslant 0.000**$	1	1	2	1	1	2
$F_{observed}$:	334							
$F_{expected}$:	75.8	$p \leqslant 0.000**$	2	1	2	1	1	2
$F_{observed}$:	322							
$F_{expected}$:	152.2	$p < 0.003$	2	2	2	2	2	2
$F_{observed}$:	571							

Note. N, norm deviation; INT, Intention; INJ, injury; 1, low level; 2, high level. Data from Löschper et al. (1984).

jective pattern of the three criteria are frequent. The remaining two remarkably frequent configurations contain subjective interpretations deviating from the objective combination. Both consist of a high norm deviation combined with low intent on the objective side, resulting in an assimilation of the high norm deviation towards the low intent on the subjective side. (In additional analyses the same pattern emerged; high norm deviation in combination with low intent on the objective side becomes low norm deviation in the subjective representation.)

Therefore it can be concluded that the frequency of a correct subjective understanding of combinations of the three definition criteria – norm deviation, intent, and injury – varies according to specific combinations and levels of the variables. "Reinterpretations" of episodes or deviating subjective representations occur for those episodes in which norm deviation and intent are varied on different levels, i.e., an inappropriate behavior is carried out without the actor's intent to harm. The analysis of subjective representations (Table 5.3) shows, moreover, that if an actor's intent is beyond doubt but the behavior is only slightly inappropriate, subjects tend to interpretate the low norm deviation as high. It can be concluded that people tend to assimilate either norm deviation or intent at respectively matching levels. Scientific conceptualizations, e.g., concerning the plausibility of a norm deviation combined with a lack of intent in operationalizing arbitrariness, seem to be influenced by the preferred "matching" interpretation of norm deviation and intent as well as everyday conceptions of combinations of the two variables as shown by our subjects.

Therefore the question arises of whether the assumption of the variables, norm deviation, intent, and injury, being independent definition criteria has any meaning at all, because some of the specific combinations could be psychologically meaningless.

To investigate this, the cross-tabulation table with all "objective situations" by all "subjective situations" was analyzed concerning the subjective representations actually deviating from the objective pattern (cf. Table 5.3).

Above all, the cross-tabulation shows that all experimental conditions or combinations of the three variables exist also in the subjective representation. Even

Table 5.3. Cross tabulation table of objective conditions and the subjective representation for norm deviation, intent, and injury in levels high and low

Count row % col % tot %	Subjective representation[a]								Row total
	222	221	212	122	211	121	112	111	
222	571	70	51	56	21	19	15	21	824
	69.3	8.5	6.2	6.8	2.5	2.3	1.8	2.5	12.4
	32.0	6.1	14.1	10.9	6.4	2.7	1.5	2.4	
	8.6	1.1	0.8	0.8	0.3	0.3	0.2	0.3	
221	296	308	18	35	25	92	9	49	832
	35.6	37.0	2.2	4.2	3.0	11.1	1.1	5.9	12.5
	16.6	26.9	5.0	6.8	7.6	12.8	1.0	5.6	
	4.5	4.6	0.3	0.5	0.4	1.4	0.1	0.7	
212	114	20	104	81	22	25	322	143	831
	13.7	2.4	12.5	9.7	2.6	3.0	38.7	17.2	12.5
	6.4	1.7	28.8	15.7	6.7	3.5	35.3	16.3	
	1.7	0.3	1.6	1.2	0.3	0.4	4.9	2.2	
122	318	141	35	143	26	61	58	50	832
	38.2	16.9	4.2	17.2	3.1	7.3	7.0	6.0	12.5
	17.8	12.3	9.7	27.7	7.9	8.5	6.4	5.7	
	4.3	2.1	0.5	2.2	0.4	0.9	0.9	0.8	
211	94	95	62	38	73	76	144	248	830
	11.3	11.4	7.5	4.6	8.8	9.2	17.3	29.9	12.5
	5.3	8.3	17.2	7.4	22.3	10.6	15.8	28.3	
	1.4	1.4	0.9	0.6	1.1	1.1	2.2	3.7	
121	175	274	9	85	27	187	8	61	826
	21.2	33.2	1.1	10.3	3.3	22.6	1.0	7.4	12.4
	9.8	23.9	2.5	16.5	8.2	26.1	0.9	7.0	
	2.6	4.1	0.1	1.3	0.4	2.8	0.1	0.9	
112	178	44	65	67	14	71	334	58	831
	21.4	5.3	7.8	8.1	1.7	8.5	40.2	7.0	12.5
	10.0	3.8	18.0	13.0	4.3	9.9	36.7	6.6	
	2.7	0.7	1.0	1.0	0.2	1.1	5.0	0.9	
111	36	194	17	11	120	185	21	246	830
	4.3	23.4	2.0	1.3	14.5	22.3	2.5	29.6	12.5
	2.0	16.9	4.7	2.1	36.6	25.8	2.3	28.1	
	0.5	2.9	0.3	0.2	1.8	2.8	0.3	3.7	
Column total	1782	1146	361	516	328	716	911	876	6636
	26.9	17.3	5.4	7.8	4.9	10.8	13.7	13.2	100.0

Objective condition[a]

[a] 1, low; 2, high; the first digit refers to norm deviation, second to intent, third to injury. Data from Löschper et al. (1984).

conditions in which norm deviation and intent are given on different levels on the objective side and which were not frequently interpreted correctly according to the configurational frequency analyses *do exist as a subjective situation,* i.e., are constructed or "fabricated" voluntarily out of other combinations. Thus the assumption that specific combinations of the definition criteria are useless is disproved. Norm deviation and intent, for example, do indeed correlate highly, but are not identical. Though there were preferential interpretations of the combinations of the experimental variables in the manner described above, *all combinations* of the single definition criteria, norm deviation, intent, and injury, are psychologically meaningful and are used in the subjective representations.

To sum up, the definition criteria – norm deviation, intent to harm, and actual injury – are important and essential for the definition of actions and situations in aggressive interactions. Compared with the perceived inappropriateness of a behavior and the actor's supposed intent to injure the victim, the form and amount of the actually occurring harm is less important for identifying the critical act as aggressive, but it does influence the selection of possible responses, i.e., judgments of the behavior as deserving punishment. The negative norm deviation and the intent to harm are necessary and, in combination, sufficient conditions for evaluating critical behavior as aggressive and punishment being held suitable. Conceptions about the further progress of the interaction, especially about the appropriateness of sanctions against the actor, were found to be more sensitive to the variation of the three definition criteria than the definition of acts as aggressive. In former studies, the influence of the definition criteria was also analyzed, but in most cases only concerning one criterion or confounding factors. This could be due to the fact that, as shown by an analysis of the subjective representations, some factors are close in their subjective meaning, i.e., are frequently interpreted in agreement with one another. Especially the negative norm deviation and the intent to injure are used in a highly correlated way; however, the two criteria are not identical though being similarly conducive for the interpretation of and reactions to the critical behavior. In general it is not the "objective" variation and combination of the definition criteria, i.e., the objective situation but rather the subjective representation of information about a critical event that determines the interpretation of the behavior and the selection of possible reactions.

The Context of Aggressive Interactions: Taxonomy of Social Situations

The conceptualization of aggressive interactions resulting from mutually performed evaluations and definitions of behavioral acts raises the question of possible information sources of these judgmental processes. There can be no doubt that people do not only take into account what happens "circumjacent" (cf. Barker, 1968, p. 19) to the critical act when evaluating it, but they also need information about the embedding situation. For example: A person takes into account the information that the crowded subway train (s)he is standing in just started when his/her neighbor stepped on his/her foot. So he/she has no reason to attribute a bad intention to harm him/her and does not label his/her act aggressive. If, on the contrary, the person is debating controversialy with another person standing in front of him/her and the other one steps on his/her foot in a similar way, (s)he may have reason to place quite a different label on the behavior. In this case, (s)he probably feels justified to react, e.g., push the other person away by hitting, his/her chest.

Thus, rather than the behavior, it is the context which determines the meaning (Gergen, 1980; Tedeschi & Lindskold, 1976). Within this concept, it is impossible to look at aggressive interactions in a vacuum (cf. Tajfel, 1972; Weinstein, 1969) because the judging persons have no criteria for defining the act concerning its appropriateness, its intent, and its harm-doing character to the victim. Instead we assume that characteristics of the social and physical environment as aspects of an embedding situation are accentuated by judges and by this are brought into a (psycho-)logical relationship with the particular action.

Stepping forward from act to interaction, within the presented concept the link between the embedding situation and the sequence of interaction is given by mutually performed processes of giving meaning to critical acts according to aspects of the context. It is noteworthy that the individual perspective on the situation is not arbitrary, so that as many different subjective representations of a situation result as persons are present. It is our very endeavour to evaluate consensual variabilities in judging critical acts dependent upon different situations. Thus it becomes obvious that norms and rules actualized by certain situations enable persons to come to a *socially accepted* behavior interpretation.

Within the presented theoretical framework, we have to look for information about the influence factors of the surrounding sociophysical situation on judgmental processes. Although the research upon the influence of the social and physical environment on aggressive behavior points in a quite different direction (by assuming a direct or indirect influence of environmental variables on individual aggressive behavior), we accept a strong impulse from the critical and summarizing literature (cf. Baron, 1977; O'Neal & McDonald, 1976; Stokols, 1977): In order to heighten ecological validity for future studies, it seems to be more appropriate and promising to analyze complex situations instead of isolating variables for research purposes.

The limitation in studying the variability of judgments and definitions of critical acts dependent upon different situations on a socially and normatively defined setting allows us, by gaining information about commonsense objectives, to differentiate situations like Barker's (1968, 1979) "behavior settings" and Stokols' (1980) "places." Depending upon the criteria used for classification, a larger or smaller number of situations result, as environmental psychology studies show. In the present study, however, we are not interested in a global description of a system but in a *differentiation of situations specific for a certain category of interactions, i.e., aggressive interactions.*

Studies on the differentiation of situations provide certain methods (e.g., Frederiksen, 1972; Moos, 1974; Price, 1974; Price & Bouffard, 1974; Price & Moos, 1975). These studies, however, use merely appropriateness judgments concerning the congruence of situation and behavior to evaluate similarities between situations in order to arrive at a classification or taxonomy of situations. The present study gains a classification using the mentioned criteria ("norm deviation" or "inappropriateness," "intent," "injury," and "aggression") to evaluate critical acts within certain situations. Thus the criteria for classifying field-specific situations are *determined by the specific form of interactions* we are interested in. In other words, the study strives for a *behavior-specific taxonomy* of situations of a specific field which is multidimensional, insofar as the classification is based on judgments of critical acts performed within these situations, using the four specific criteria for evaluating behavioral acts as aggressive.

The exact methodological procedure was performed in several steps.

1. Identification of "Neuralgic Places." In the first study, 105 13- to 16-year-old pupils were asked to identify specific situations in which interactions they defined as aggressive occurred either remarkably often or seldom. In this manner, a kind of map of neuralgic places of the field could be elaborated.

2. Identifications of Descriptive Dimensions. One can determine dimensions or variables by which social situations in the field can be described when considering aggressive interactions in three ways: the results from the first study, the environmental psychology of aggression, and studies of specific problems of the field. The resulting variables are not general but are already related to the object in question, i.e., aggressive interactions. Table 5.4 shows the selected variables, together with two selected extreme levels each.

In this stage of research it can only be said that these field-specific variables have some relation to aggressive interactions. We do not yet know the exact connection. This empirical step, however, is unmistakably necessary in order to supply evidence for the selection of variables which are determined by the realities of the field, both in this stage of research and for the further steps.

If we take just two possible levels of the selected dimensions as given in

Table 5.4. Variables characterizing social situations in schools with two extreme levels each

Variable	Levels
Social density (D): number of persons together in a particular place	High (+) / low (−)
Spatial mobility (M): persons keeping quiet or moving around in the place	High (+) / low (−)
Audience (A): bystanders noticing the critical interaction	Yes (+) / no (−)
Presence of teachers (T): teacher or janitor nearby or not	Yes (+) / no (−)
Achievement pressure (P): particular performance required or not	Yes (+) / no (−)
Stress (S): participants tired from previous activity or relaxed and rested	Yes (+) / no (−)

Table from Linneweber, Mummendey, Bornewasser, & Löschper (1984).

Table 5.4 and list all the theoretically possible combinations of the variables, we get $2^6 = 64$ descriptions of situations by six variables each.

3. Acquiring Written Descriptions of Situations and Critical Interactions. Given a characterization of a situation by certain levels of each variable (e. g., D +, M +, A +, T −, P +, S −), it is possible to formulate a verbal description of a concrete situation which everybody recognizes, e. g., "Art class, first lesson of the day. The teacher is in the room next door. All children have to fetch water in order to mix their paints. Thus there is a lot of running to and fro in the art room. Everybody notices that . . ."

In this situational context a critical act can be embedded: ". . . Jürgen has knocked Ulrich's water pot out of his hand." Taking all the 64 descriptions and formulating written situations within which critical interactions happen, we see that all of the 64 theoretically possible combinations are also meaningful in a real context. Selecting two forms of critical interactions (verbal and physical), we altogether have 128 descriptions of situations and interactions.

4. Studying Judgments of the Episodes on the Aggression-Specific Dimensions. About 700 school children aged between 14 and 19 judged critical interactions, verbally described within the specific situative context. Every subject rated only eight episodes. The critical act varied in form, verbal or physical, and in type. A pretest ensured that the interactions were judged equally as medium aggressive, irrespective of any social context (in the pretest subjects had no information about the context of the situation). For this kind of "moderate aggressive" act, a distinct variability in different contexts was expected.

5. Grouping Variables. The dependent variables were the previously given definition criteria.

All variables were operationalized by means of a bipolar rating scale with seven graduations. The scale nomenclature is the same as in the previously reported study.

6. Classification of Situations. The situations were classified by patterns of judgments in order to construct a field- and behavior-specific taxonomy of situations. By cluster analysis (computed according to rating means of the critical event per situation on four judgmental dimensions), groups of situations were evaluated in which similar judgments of critical events resulted.

Figure 5.5 shows the dendrogram for situations with only physical interactions. The graph of intercluster distance or degree of generalization suggests a three-cluster solution for situations with physical interactions.

Figure 5.6 shows that each cluster is characterized by rating means on the grouping variables; this means that in each cluster the critical act is judged in a specific way using the four dimensions. The figure shows a conspicuous variation in rating means between clusters on the dimension "intent". Therefore, intent can be considered as the group discriminating main variable for physical interactions.

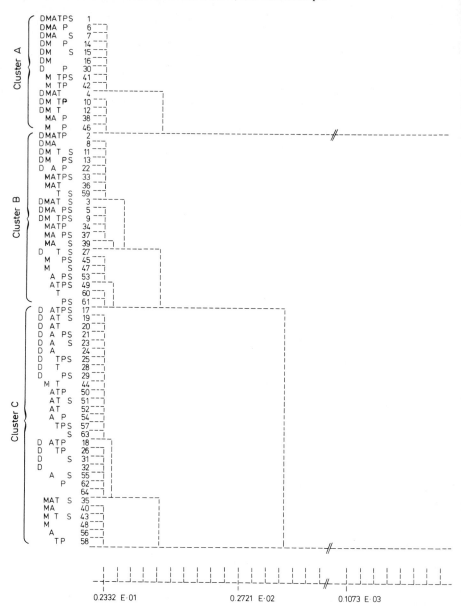

Fig. 5.5. Dendrogram for 64 objects (situations containing physical interactions). A printed grammalogue (DMATPS) indicates a high variable value on the respective description dimension. Data from Linneweber et al. (1984)

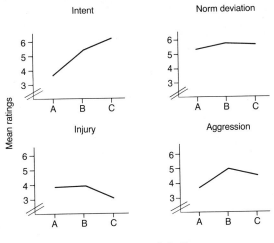

		Cluster A	Cluster B	Cluster C	Total
Intent	Mean	3.63	5.34	6.21	5.36
	SD	0.59	0.55	0.21	1.09
Injury	Mean	3.83	3.95	3.03	3.51
	SD	0.56	0.38	0.40	0.62
Inappropriate	Mean	5.33	5.75	5.63	5.60
	SD	0.27	0.36	0.27	0.34
Aggressive	Mean	3.69	4.95	4.47	4.46
	SD	0.39	0.46	0.43	0.63

Fig. 5.6. Description of clusters A, B, C by rating means on the judgment dimensions. Physical interaction situations. Data from Linneweber et al. (1984)

Figure 5.7 shows the dendrogram for situations with verbal interactions. Here the five-cluster solution appears to be appropriate.

Figure 5.8 presents the computed characterizations of the clusters for situations with verbal interactions. "Injury" tends to be the group discriminating main variable.

At this state of computation, the interpretation is possible that in the field, "school" situations can be differentiated according to judgments of critical behavioral acts occurring within them. For each group of situations, characterizing means can be computed which describe group-specific judgments of critical acts happening in situations collected in the cluster. Intercluster differences for situations with physical interactions mainly depend upon judgments of intent; intercluster differences for situations with verbal interactions mainly depend upon judgments of injury.

The design of the present study allows the following statistical procedure. The situations to be classified were systematically constructed from certain constellations of the description dimensions and were thereby comparable by means of quasi-independent variables. It becomes possible then to take an evaluative step connected to the cluster analysis. This is done by a configurational frequency

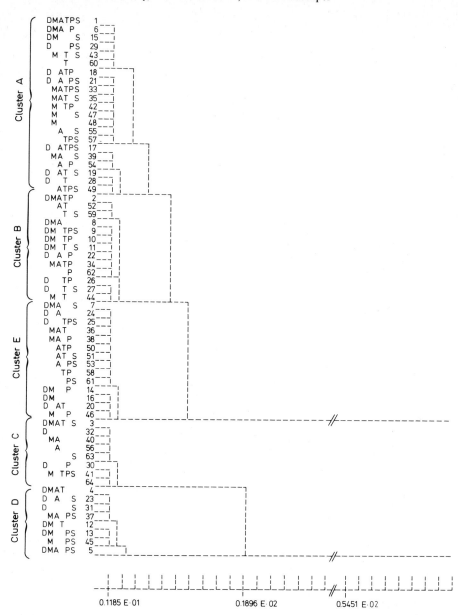

Fig. 5.7. Dendrogram for 64 objects (situations containing verbal interactions). A printed grammalogue (DMATPS) indicates a high variable value on the respective description dimensions

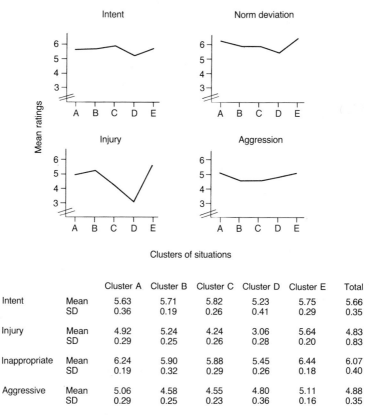

Fig. 5.8. Description of clusters A, B, C, D, E by rating means on the judgment dimensions. Verbal interaction situations

		Cluster A	Cluster B	Cluster C	Cluster D	Cluster E	Total
Intent	Mean	5.63	5.71	5.82	5.23	5.75	5.66
	SD	0.36	0.19	0.26	0.41	0.29	0.35
Injury	Mean	4.92	5.24	4.24	3.06	5.64	4.83
	SD	0.29	0.25	0.26	0.28	0.20	0.83
Inappropriate	Mean	6.24	5.90	5.88	5.45	6.44	6.07
	SD	0.19	0.32	0.29	0.26	0.18	0.40
Aggressive	Mean	5.06	4.58	4.55	4.80	5.11	4.88
	SD	0.29	0.25	0.23	0.36	0.16	0.35

analysis (Bartoszyk & Lienert, 1978, Krauth & Lienert, 1973; Lienert & Krauth, 1975), which allows a statistically validated description of the resulting clusters by some kind of "passive variables", which means that they were not used to compute the classification.

Conspicuously frequent configurations of situations are identified in the particular clusters. For the interpretation of each cluster, a "modal situation" can be constructed (cf. Table 5.5). With regard to the dependent variables, i.e., cluster-typical judgment patterns, corresponding conspicuously frequent combinations of levels of the independent variables in each cluster can be found and aggregated in a cluster-typical situation to represent it. Thus the interpretation can step forward from these typical situations and look at similarities or differences between clusters and compare these with the characterizations of clusters on grouping variables shown in Figures 5.6 and 5.8. Looking at the selected field, interpretations are possible concerning variations of evaluating acts as aggressive, depending upon the surrounding situational context. (For more detail cf. Linneweber, 1981; Linneweber, Mummendey, Bornewasser, & Löschper, 1984).

Table 5.5. Modal situations characterizing clusters

Physical interactions	Verbal interactions
Cluster	Cluster
A: D+ M+ A− T− P+ S−	A: A+ T+ S+
B: D− M+ T− P+ S+	B: D+ M+ A− T+
C: M− A+ P−	C: D− M− A− T− P− S−
	D: M+ T− P+ S+
	E: not characterizable

Note. D, density; M, mobility; A, audience; T, presence of teachers; P, achievement pressure; S, stress.

The results confirm that identical behavioral acts may be evaluated differently, depending upon the respective surrounding situational context. Furthermore, situations can be reliably classified using dimensions that lead to a behavior- and field-specific taxonomy of situations. For further studies, the present inquiry provides a well-grounded restriction of situations to be analyzed, for which the modal situations can be drawn up, the more so as their characterization has been worked out on the judgment dimensions. Such a meaningful limited number of situations enables further studies, e. g., by the choice of an audiovisual medium for presenting the situation sketches with the goal of approximating natural situations.

The Actor-Victim Divergence in Aggressive Interactions

Although there are several hints from everyday life as well as scientific literature on aggression (Da Gloria & De Ridder, 1977; Felson, 1978; Tedeschi et al. 1974) of a divergence between actor and victim in aggressive interactions, evaluating the actor's behavior against the recipient as function of their interaction-specific position, this assumption has not yet been tested empirically. Several of our own experiments were done to this purpose.

In these studies we did not use actual actors and victims of real life aggression, but subjects who were instructed to take over empathically the perspective of an actor or a victim whose interaction is depicted in a short videotape. These videotaped episodes and the instructions were carefully produced, and manipulation checks proved the procedure was successful (cf. Mummendey, Linneweber, & Löschper, 1984). The presented *episodes* took place in a natural field, i. e., schools. They were produced on the basis of verbal reports by pupils of typical, field-specific aggressive interactions. In all scenes, the specific context of the critical behavior, e. g., a school yard or a classroom, was shown first. The main characters (Michael and Thomas) were then shown having a relatively harmless argument (e. g., pushing or shouting at each other) with other pupils around them. Careful attention was paid not to present one of the two interactants as the unequivocal initiator of the argument or clear aggressor. After this the critical interaction was presented. It consisted of two interrelated, nearly identical events: in the first part (segment I) of the interaction, one person – the actor and initiator – performed a verbal or physical attack directed against the other. The resulting injury for the victim was depicted and after a short intermezzo signaling that some time had passed, the victim performed a similar, nearly identical attack against the former actor, who was then the new victim. Thus the victim of segment I became actor in segment II and, therefore, re-actor. After this the same amount of injury was depicted as in segment I. Five different videotaped episodes were used. Each subject rated one randomly chosen scene.

Special attention also was paid to the *instruction to take over the perspective* of one of the two interacting persons. In order not to present different films to the group of actors vs victims produced out of different points of view or varying the salience of the two persons, the judgment material was always identical for the experimental conditions but was preceded by a presentation of either the actor or the victim. This videotaped presentation of the respective stimulus person, i.e., showing him/her working alone in a library, emphasized the similarity and propinquity of the presented person and the subject and presented an opportunity to become acquainted with him/her.

The verbal instruction (soundtrack of the videotape) impressively demanded empathy with the respective stimulus person, imaging as vividly as possible his/her feelings and emotions and seeing everything with his/her eyes. The manipulation checks showed that most of the subjects were able to take over the respective perspective in the desired way and judged the critical episode out of it (for more detail cf. Mummendey et al., 1984).

Therefore, all evaluations and judgments of critical behavior in our experiments were done by subjects "vicariously," i.e., out of the perspective of the stimulus person.

The experiments tested several aspects or specifications of the general "position-specific divergence" assumption. It is assumed that the typical relation between actor and victim in aggressive interactions is a conflict, i.e., the victim judges the actor's behavior directed against him/her as inappropriate, according to situation-specific rules, norms, and expectations of adequate behavior; on the contrary, the actor him-/herself evaluates his/her own behavior compared with other alternatives available under the particular circumstances.

Actor and Victim Judging – A Single Critical Behavior

This hypothesis was tested by the following study: In one experimental group subjects were instructed to take over the actor's perspective, while the other group took over the perspective of the victim. Out of both perspectives (or interaction-specific positions) the critical behavior presented in a short videotape was evaluated as to its situational appropriateness and as to whether it was aggressive. In addition to the variable "perspective", a second one, "segment", was introduced. The subjects rated either a *single critical act* not preceded by a provocation or followed by a revenge (the presentation of the videotape was finished in this condition after segment I) or an identical act as the first segment or critical

Table 5.6. Mean ratings of inappropriateness (1) and aggression (2) ($N = 81$)

Perspective		Segment		
		Single event		First of two events
Actor	1	3.20 } $n = 10$		3.86 } $n = 30$
	2	2.30		3.33
Recipient	1	4.26 } $n = 8$		5.24 } $n = 33$
	2	2.50		2.50

Note. High values for appropriateness mean evaluations as inappropriate; low values for aggression mean high aggressiveness. Analysis of variance (ANOVA) of these scores revealed only a significant main effect of perspective on appropriateness scores ($F (1/77) = 13.445$, $p < 0.0005 = <1\%$, alpha adjusted). Data from Mummendey, Linneweber, & Löschper (1984).

interaction of mutually occuring critical behavior (both segments of the interaction were presented in the videotape and the subjects were instructed to rate the first one).

Thus a 2×2 factorial design varying "perspective" (actor vs victim) and "segment" (single event vs first of two events) for individual groups was used. Judgments of the actor's behavior concerning situational appropriateness and labeling as aggressive on seven-point bipolar scales were demanded. Eighty-one male pupils, familiar with the depicted interactions and of same sex, age, and grade as the main characters in the episodes participated in the study as subjects. The results for both dependent variables are presented in Table 5.6.

The main effect of "perspective" and individual comparisons show that subjects from the actor's persepctive evaluate the actor's critical behavior as more appropriate and are less likely to label it as aggressive than subjects in the condition recipient. This actor-victim divergence in evaluating the appropriateness of the actor's behavior (and, in turn, its definition as aggressive) emerges irrespective of the critical act being a single hostile act or just the first segment of a longer course of an aggressive interaction.

This divergent evaluation paves the way for the course of the following interaction: the victim feels justified in retaliating and in punishing, in an "appropriate" way, the initiator's behavior evaluated as inappropriate and aggressive. Thus when choosing an adequate re-action, the former victim (now the actor in segment II of the interaction) evaluates his behavior as justified, situationally suitable, and more appropriate than the former behavior directed against him.

Thus it is assumed that actor and victim in each segment of an aggressive interaction differ in judging the situational appropriateness and in defining the respective actor's behavior as aggressive.

Evaluations of Aggressive Actions and Hostile Responses

The experiment described above supports the assumption of the typical actor-victim divergence concerning a single critical act, as well as the first of two interrelated ones.

In two experiments reported below, a related assumption was tested: whether this divergence results irrespective of the critical act being an initiative action (first segment) or a re-action (second segment of the interaction). Therefore, the evaluation of the initiative action and the reaction were compared. For this purpose subjects in one experimental group were instructed to take over the perspective of the actor in the first event ("initiator"). Subjects in the other group were demanded to identify with the victim in the first event and actor in the second segment ("re-actor"). In the first of these two experiments concerning action and re-action, one experimental group in both "positions" (initiator vs re-actor) evaluated only the first critical event or "segment" (according to the results of the experiment reported above, the first segment of the interaction was substituted by presentation of the single event, i. e., the presentation of the videotape was finished after the first event), while the other group judged the second critical event (both events were presented with the instruction to judge the second segment, the re-action).

In the second experiment concerning action and re-action, a design with repeated measurements was used instead, i. e., out of both "positions" (initiator vs

Table 5.7. Mean ratings of inappropriateness (1) and aggression (2) ($N=81$)

Position		Segment	
		Single event (action)	Second of two events (re-action)
Initiator	1	3.20^b $\}$ $n=10$	4.63^a $\}$ $n=30$
	2	2.30^c	2.97^c
Re-actor	1	4.62^{ab} $\}$ $n=8$	3.82^{ab} $\}$ $n=33$
	2	2.50^c	2.97^c

Note. High values for appropriateness mean high inappropriateness; low values for aggression mean high aggressiveness. Cells having different superscripts are significantly different at less than the 0.05 level of confidence according to the Scheffé test. Superscripts *a* and *b* refer to the first dependent variable and *c* to the second dependent variable.
ANOVA of these scores revealed only a near-significant effect ($F (1/77)=4.897$, $p<0.03 = <15\%$, alpha adjusted) for appropriateness scores. Data from Mummendey et al. (1984).

Table 5.8. Mean ratings of appropriateness (1) and aggression (2) ($N=63$)

Position		Segment	
		Action	Re-action
Initiator	1	3.83^a	4.63^{ab} $\}$ $n=30$
	2	3.33^c	2.97^{cd}
Re-actor	1	5.24^b	3.82^a $\}$ $n=33$
	2	2.50^d	2.97^{cd}

Note. Cells having different superscripts are significantly different at less than the 0.05 level of confidence according to the Scheffé test. ANOVA with repeated measures on "segment" revealed a significant effect (F $(1/61)=9.79$, $p<0.0027=1\%$, alpha adjusted) for appropriateness scores. Data from Mummendey et al. (1984).

re-actor) the first *and* the second segment of the interaction (action and re-action) was evaluated concerning its situational appropriateness and definition as aggressive.

A 2 × 2 factorial design was used in both experiments to test the influence of the variables "position" (initiator vs re-actor) and "segment" (single event vs second event and first event and second event) on the dependent variables of appropriateness and aggression, measured with seven-point bipolar rating scales. For both designs an interaction effect of "position" and "segment" was expected: while in the first event ("action"), the initiator (actor here) evaluates his behavior as more appropriate and less aggressive than the re-actor (victim here), in the second segment ("re-action"), the initiator (victim here) judges the actor's behavior as more inappropriate and aggressive than the re-actor (actor here).

The results of both experiments confirm the assumption on the whole (cf. Tables 5.7 and 5.8). The individual comparisons in the experiment with individual groups show that for the first segment, i.e., the action, the assumed divergence between initiator and reactor in evaluating the situational appropriateness emerges. This is not true for the second segment, though the means tend in the expected direction. The divergence could not be shown for the aggression ratings. Moreover, initiators evaluate their own hostile initiative behavior, i.e., the action, as even less inappropriate than the actor's identical re-action (cf. Figure 5.9). In the repeated measurement design, for the action the assumption of the actor-victim divergence is confirmed; for re-action, means tend in the expected direction.

While in the first segment of the interaction, the re-actor (victim here) evaluated the actor's behavior as more inappropriate and more aggressive than the initiator himself, in the second segment (re-action) the initiator judged the actor's behavior as more inappropriate, but not significantly more aggressive than the re-actor. Similar to the subjects evaluating the critical interaction out of the ini-

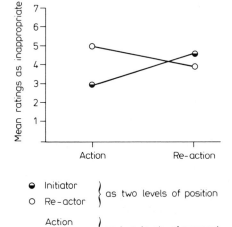

Fig. 5.9. Mean ratings of behavior as inappropriate in conditions "position" (initiator vs re-actor) and "segment" (action vs re-action). Independent groups. Data from Mummendey et al. (1984)

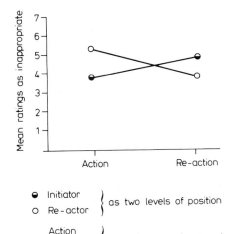

Fig. 5.10. Mean ratings of behavior as inappropriate in conditions "position" (initiator vs re-actor) and "segment" (action vs re-action). Repeated measurement. Data from Mummendey et al. (1984)

tiator's perspective in the former experiment, here the re-actors judge their own hostile behavior as more appropriate than the other's initiative action (cf. Figures 5.9 and 5.10).

It can be concluded that the divergence between actor and victim in aggressive interactions concerning the situational appropriateness (and, in turn, the evaluation as aggressive) of the respective actors' behavior emerges even irrespective of the critical event being an initiative hostile action or a hostile response. These results can be regarded as a supplement to results already known concerning the different evaluation of action and re-action in aggressive interactions (cf. Da Gloria & De Ridder, 1977; Felson, 1978; Tedeschi et al., 1974), where judgments out of the observer's perspective were evaluated. Moreover the results seem to show that one's own behavior (initiators in the first, reactors in the second design) is evaluated as more appropriate than the other's behavior.

The Self – Other Divergence in Evaluating Aggressive Behavior

In all the experiments described above, to compare evaluations of the actor's behavior out of different perspectives or positions it was necessary that subjects evaluate differently their own (the actor judges actor's behavior) or the other's behavior (the victim evaluates the actor's behavior).

In accordance with results obtained by the self-other paradigm in the attribution literature (cf. Zuckerman, 1979), it is assumed as a further specification of the hypothesis of the actor-victim divergence that, irrespective of the course of the interaction, i.e., the critical behavior being action or re-action, and the positions held by the judges, i.e., initiator or re-actor, one's *own* behavior is evaluated as less inappropriate and aggressive as compared with the *other's* behavior.

This assumption was tested in a 2 × 2 factorial design varying "Position" (ini-

Table 5.9. Mean ratings of appropriateness (1) and aggression (2) ($N = 63$)

Person		Position			
		Initiator		Re-actor	
Own behavior	1	3.83		3.82	
	2	3.33	$n = 30$	2.97	$n = 33$
Other's behavior	1	4.63		5.24	
	2	2.97		2.50	

Note. ANOVA with repeated measure on "person" yielded only a significant main effect for person for the appropriateness scores ($F(1/61) = 10.08$, $p < 0.0024 = < 1\%$, alpha adjusted). Data from Mummendey et al. (1984).

tiator vs re-actor) and "person" (own behavior is judged and other's behavior is judged) with repeated measurements on the second variable. The results for the dependent measures appropriateness and aggression ratings are presented in Table 5.9.

The only resulting main effect of "person" confirms the hypothesis that one's own behavior is judged as more appropriate than the similar or identical behavior of the other person involved. This results irrespective of whether the person holds the position of initiator and reactor and thus whether one's own behavior takes place as an initiative action or a hostile reaction.

The presented results confirm several aspects of the general assumption of an actor-victim divergence in judging appropriateness as a crucial characteristic of aggressive interactions.

Aggressive interactions are characterized by two persons, being at a certain point of time in the interaction-specific positions of actor and victim, who differ in evaluating the actor's critical behavior as normatively appropriate and aggressive. This divergence is something like the power transmission for aggressive escalations. The respective recipient feels bound and justified to respond to his opponent's aggressive and inappropriate behavior in an "appropriate" way. For this hostile re-action, the actor-victim divergence again arises as each person relates the critical behavior to situationally relevant norms and strikes the balance with the previous course of the argument. Moreover, one's own critical behavior is always judged as more appropriate and less aggressive in tendency than the opponent's behavior. Courses of aggressive interactions thus can be considered as chains of segments in which the respective positions of actor and victim and the corresponding evaluations of the critical behavior change.

Conceptions About the Progress of Aggressive Interactions

In order to determine regularities in the course of an aggressive interaction, social-consensual conceptions of typical connections between the judgment of a critical behavior and the following reaction to it should be explored. To this purpose, regularities concerning the relation of information about three elements of an episode were evaluated: (1) information about the critical action, (2) mediatory information concerning the connection between action and reaction, and (3) information about the reaction.

By configuration frequency analytical categorization of progress curves, typical configurations of the variables selected by the subjects were determined (for more detail cf. Mummendey, Löschper, Linneweber, & Bornewasser, 1984).

Information about the first critical action and about the reaction were systematically varied. In several verbal episodes a critical action was described to which one of three different reactions followed, i.e., *escalation, termination,* or *compensation by the harm-doer.* These three courses of aggressive interactions can be looked upon as interesting in aggression research. Subjects had to choose between the alternatives of "connecting sections," i.e., mediatory information mentioned above, to combine the beginning and the end to form a complete, plausible, and meaningful episode. To this purpose four different aspects were dichotomously presented as mediatory information:

- Information about what had happened previously (a hostile attack toward the present attacker or not)
- Information about the personal standard of the victim concerning the appropriateness of performing aggressive actions (yes or no)
- Information about the amount of harm done to the victim (high or low)
- Information about the probability of negative consequences the victim would expect when reacting aggressively to the attacker (high or low)

The results show a very impressive unanimity in constructing plausible configurations of the informations into a meaningful episode. The choice of the mediating information was mainly influenced by the kind of reaction presented, i.e., termination, escalation, or compensation. Thus, there wasn't any difficulty for the subjects to imagine combinations of identical aggressive actions with these complementary reactions, and they chose consensually agreeing combinations of the mediating information to make these combinations plausible. Apparently persons use definite and rather homogeneous conceptions about the information or conditions which may give sense either to one or the completely opposite reaction to a particular hostile action.

Conclusion

The purpose of this chapter was to present a conceptualization of aggression as a specific kind of interaction that means to offer a social-psychological perspective on aggression research. Several studies performed within this frame of reference were presented to demonstrate some possibilities for empirical work on propositions of this social-psychological origin. Although still preliminary and only providing first empirical steps, the results support the decision for proceeding in this direction.

There seems to be sufficient evidence for the use of an interactional, instead of an individual, concept of aggression. The studies clearly show a large variability in evaluating identical individual actions as aggressive or not, depending upon the systematic variation of specific contextual and interactional conditions. The postulated divergence of appropriateness evaluations between actor and recipient turn out to be obvious. Especially this divergence, which means a *special relation* between the judgments of two individuals or parties, represents an *interactional term,* a characteristic of what is *between* the two participants. Further theoretical and empirical work could be concerned with the more fundamental processes that result in this position-specific dissent. It would be of interest to know some more about the process of generating such an evaluation, about the different kinds of information used to form appropriateness judgments (for more details about the role of moral judgments see Rule & Ferguson, this volume).

One crucial point for the dissent seems to be the diverging perceptions of the avoidability of the aversive stimulation. The actor may include a selection of information and arguments in his/her net balance different from those included by the recipient. The disagreement may be due to a lack of correspondence in the *selection* of the relevant items or, if the selection is correspondent, due to differences in giving weight to these items. Thus, the actor in a certain situation only realizes a limited range of behavioral alternatives, which in his/her view should have been known by the actor as well and which would not have produced the aversive effect. Or both recipient and actor realize the identical range of possible behavioral alternatives, but they disagree about the degree of necessity to choose the particular one producing the aversive effect. Analogous to these ideas, the different kinds of reactions to predicaments (proposed by Tedeschi & Riess, 1981) like excuses, justifications, etc. as impression management strategy provide explanations for an actualized or perceived dissent in order to diminish it and to come to an acceptance of an over all appropriateness of a critical event. The analysis of the differential use and *success* of various kinds of strategies, depending upon the characteristics of the type of event in question, could provide information about the essential features of the dissent.

The postulate of the perspective-specific divergence as a fundamental characteristic of aggressive interactions cannot seriously be challenged by the argument sometimes presented (e. g., Zumkley, this volume) that there can be an actor intending with full consciousness and deliberation to cause harm. More precisely, it must be stated that the existence of such an actor could be imagined by an observer (or psychologist). A theory of aggression could not, however, be founded solidly on merely the imagination of the harm intent existing within the actor, hidden from the eye of the observer. The actor may be asked about it, but may deny or admit intent; in any case nobody knows whether the answer is trustworthy.

Within the present concept, it is proposed that, in contrast to the victim, the actor will have perceived good reasons to produce these consequences. These reasons may differ in the amount of social consensus that supports them. The social consensus may range from very wide agreement, as expressed by moral obligations to one's country, religious group, etc., to only very limited consensus provided by a smaller social subgroup or even only personal reasons not really shared by anyone else. For the decision of the actor to perform a critical act, it is of minor importance whether an assumption of consensus about the appropriateness of the critical action is realistic or not. But the actor would not have chosen this action if – at the moment of choice – the pros had not dominated the contras.

Out of these considerations follows the necessity not to stop with asking only whether a specific harm or injury of the victim was the goal of an action. Moreover, the production of the victim's harm may be only a part of that goal. As actual effect, this harm is the spectacular and easily observable one. A more analytical look at the actor's goal might reveal the victim's harm only being the actor's subjectively adequate means for restoring a previously disturbed state of ease. This consideration agrees to some extent with the conception of aversively stimulated aggression (Berkowitz, 1982). For Berkowitz, too, "It is not the objective nature of the aversive incident that is important, but how aversive it is thought to be" (p. 254). In the eye of the actor, the harm infliction is a response to a previously suffered uneasiness or injury, with the perceived function of reducing the unpleasant state. This does not necessarily include the assumption of always conscious cognitive checks about the specific usefulness of this "tool" in a particular case. We may think of rather spontaneous-looking responses which from a more remote point of view seem to lack rationality, but are for the actor, under the particular circumstances, the optimum available responses.

In contrast to the approach of Berkowitz, the present approach does not end up with the statement of the relevance of individually realized aversive stimulation. Moreover, this statement presents the *beginning* of a search for the conditions which let people interpret events as aversive and feel legitimate in "paying back." We do not think that perceiving an event as aversive is a completely idiosyncratic process. We try, moreover, to take a closer look at the *social character* of perceived aversion or harm (cf. Mummendey, Bornewasser, Löschper, Linneweber, 1982). There are social uniformities within a certain social context shown by the "inhabitants" when defining something as harmful or harmless. People talk to each other about what is happening to them and how to evaluate it; they share to a certain extent common evaluations of these events. The study about the behavior-specific taxonomy of situations reported above shows some hints for the variation of these evaluations according to the variation of certain aspects of the context; these evaluation differences are shared by a large group of individuals. Studies about the influence of perceived violation of different norms upon the variation of aggressive reaction show the same direction (cf. Bornewasser, Mummendey, Löschper, & Linneweber, 1982). Further developments in theory should move, therefore, toward identification and systematization of exactly those aspects of the social context which supply its participating individuals with information about what is to be accepted and what is to be rejected. Theory about aggressive behavior – and about other kinds of social behavior as well – should overcome its limitation to the single individual as a self-sufficient behavior-producing unit of analysis.

As several authors have already stressed (Billig, 1976; Stroebe, 1979; Tajfel, 1972, 1981), human beings do not live in a social vacuum. In traditional social psychological theories there seems to be "an implicit vision of social systems as consisting of individual particles floating, as in some kind of brownian movement, within a statistically homogeneous social medium" (Tajfel, in press, p. 2). Instead of those approaches which individualize social behavior by principally limiting the conceptualizations to structures and processes internal to the individual organism, we strongly argue for conceptualizations transcending the individual as the unit of analysis. Besides modeling these individual processes, concepts about systematics in the social structure, of how these individuals are related to each other within the social context, what the "material" of relations between individuals and larger social units may be are necessary. For other areas than the present one, examples for this kind of approach were already presented (cf. theory of intergroup conflicts by Tajfel & Turner, 1979, or Hewstone & Jaspars 1982) on attribution processes, or the research on social representations (Herzlich, 1973; Moscovici, 1981).

For the area of aggression, we too try to take into account the primarily social character of this problem. Thus, we stress the necessity of leaving the concepts typical in traditional aggression research within psychology; we propose a change of the *prototype* of aggressive behavior in social psychological theory and research. Everybody who takes the risk for a while of noticing what kinds of different aggressive and violent events are happening is confronted with innumerable events of offense, torture, discrimination, expulsion, murder. And what comes to be known presents only a selection of the universe of violence presented by the respective information media. As Nagel (1978) put it "The great modern crimes are public crimes. To a degree the same can be said of the past, but the growth of political power has introduced a scale of massacre and despoliation that makes the effects of private criminals, pirates and bandits seem truly modest" (Tajfel, in press, p. 75).

The conceptualization of aggression should be prototypical for the *whole range* of cases subsumed under this phenomenon. Thus, the description of events of larger social range should be possible as well. But instead of a mere addition of a more or less large number of single, self-sufficient "behavior-producing" units, concepts must be looked for which could provide an access to questions about what makes these individuals homogeneous as actors as well as victims, what facilitates the shared perception of suffering harm, what leads them to personalize the cause of injury and, by this, define the injury as avoidable. Questions of this kind show the direction of social and historically relevant ideologies and myths about what, at a certain time within a certain social context, is just to do and just to get, what is the tolerable range when norms of justice or norms of appropriateness are violated.

Instead of extrapolating from the single-individual model of behavior to broader social or collective phenomena – and the largest number of the most serious harm-doings appear to be collective ones – we propose to do it the other way around. We argue for looking primarily at the essentials of these social phenomena, to conceptualize these as basic prototypes of aggression, and then to transfer this model to the more "private-looking" cases, as well.

One objection often articulated against such a social psychological perspective is that *individual differences* among the different actors would be neglected or even denied (cf. Zumkley and Kornadt, this volume). A closer look at the approach proposed by the present authors will reveal that this is not the case. The concept of aggression as social interaction only changes the part attributed to the individual in the manufacture of this product. The former autarky of the individual in producing aggression is now reduced only to a contribution, whereas the whole is produced by a "division of labor" among actor and recipient and outside observer. This altered position of the individual shifts the interest from former personality variables like "aggression motive" or "aggressivity" to variables which may describe different impacts upon the essential "choice points" of the interaction sequence. Individuals may differ, e.g., in perceiving the crucial definition criteria as fulfilled more or less quickly; they may tend either to personal or to situational explanations of causes of events; they may feel more or less easily attacked; or they may differ with respect to their beliefs and ideologies referring to necessities of order and the restoration of justice. They may differ with respect to their attachment to particular social groups and relationships between their own and other social groups, which could lead to differences of conceptions about which distribution of resources is just and which is not. In other words, they may rely on a different social consensus about how to evaluate certain actions and their consequences.

With the perspective outlined above, the consideration of individual differences will begin with the conceptualization of the basic interactional qualities of aggression. Thus a central question will be how the divergence between actor and recipient concerning the appropriateness of an action and its effects may arise. As mentioned above, the individual murderer chooses his/her deed – be it with more or less consciousness – because (s)he perceives it, at least at the moment of performance, as the relatively adequate one. Unfortunately for *this* aggressor, there probably won't be a broader consensus supporting his/her evaluation of adequacy. Unfortunately for millions of victims, in the majority of cases a considerable social consensus exists.

References

Arendt, H. *On violence.* New York: Harcourt, Brace & World, Inc., 1970.

Barker, R. *Ecological psychology: Concepts and methods for studying the environment of human behavior.* Stanford, Calif.: Stanford University Press, 1968.

Barker, R.G. Settings of a professional lifetime. *Journal of Personality and Social Psychology,* 1979, *37,* 2137–2157.

Baron, R.A. Environmental and situational determinants of aggression. In R.A. Baron (Ed.), *Human aggression.* New York: Plenum Press, 1977.

Bartoszyk, G.D., & Lienert, G.A. Konfigurationsanalytische Typisierung von Verlaufskurven. *Zeitschrift für experimentelle und angewandte Psychologie,* 1978, *25,* 1–9.

Berkowitz, L. Aversive conditions as stimuli to aggression. In L. Berkowitz (Ed.), *Advances in Experimental Social Psychology* (Vol 15). New York: Academic Press, 1982.

Berkowitz, L., Cochran, S., & Embree, M. Physical pain and the goal of aversively stimulated aggression. *Journal of Personality and Social Psychology,* 1981, *40,* 687–700.

Berkowitz, L., & Donnerstein, E. External validity is more than skin deep. Some answers to criticisms of laboratory experiments. *American Psychologist,* 1982, *37,* 245–257.

Bernstein, F. *Der Antisemitismus als Gruppenerscheinung.* Nachdruck der 1. Auflage Berlin, Jüdischer Verlag, 1926. Königstein/Taunus: Jüdischer Verlag im Athenaeum Verlag, 1980.

Billig, M. *Social psychology and intergroup relations.* London: Academic Press, 1976.

Bornewasser, M., Mummendey, A., Löschper, G., & Linneweber, V. Aggressive Interaktionen und normativer Kontext: Einflüsse differentieller Normverletzungen auf das Aggressionsurteil. *Zeitschrift für Entwicklungspsychologie und pädagogische Psychologie,* 1982, *14,* 308–323.

Buss. A. H. *The psychology of aggression.* New York: Wiley, 1961.

da Gloria, J., & De Ridder, R. Aggression in dyadic interaction. *European Journal of Social Psychology,* 1977, *7,* 189–219.

Epstein, S., & Taylor, S. P. Instigation to aggression as a function of defeat and perceived aggressive intent of the opponent. *Journal of Personality,* 1967, *37,* 265–289.

Felson, R. B. Aggression as impression management. *Social Psychology,* 1978, *41,* 205–213.

Ferguson, T. J., & Rule, B. G. An attributional perspective on anger and aggression. In R. Geen & E. Donnerstein (Eds.), *Aggression: Theoretical and empirical reviews* (Vol. 1): *Method and theory.* New York: Academic Press, 1983.

Forgas, J. P. Episode cognition: Internal representations of interaction routines. In L. Berkowitz (Ed.), *Advances in Experimental Social Psychology* (Vol. 15). New York: Academic Press, 1982.

Frederiksen, N. Toward a taxonomy of situations. *American Psychologist,* 1972, *27,* 114–123.

Gergen, K. J. Towards intellectual audacity in social psychology. In R. Gilmour & S. Duck (Eds.), *The development of social psychology.* New York: Academic Press, 1980.

Graumann, C. F. Interaktion und Kommunikation. In C. F. Graumann (Ed.), *Handbuch der Psychologie* (Band 7): *Sozialspsychologie, 2.* Halbband, 1109–1262. Göttingen: Hogrefe, 1972.

Graumann, C. F. Die Scheu des Psychologen vor der Interaktion. Ein Schisma und seine Geschichte. *Zeitschrift für Sozialpsychologie,* 1979, 10, 284–304.

Harris, B., & Harvey, J. H. Attribution theory: From phenomenal causality to the intuitive social scientist and beyond. In C. C. Antaki (Ed.), *The psychology of ordinary explanations of social behaviour.* London: Academic Press, 1981.

Herrmann, T. *Handlungstheoretische Aspekte der Aggression.* In H. Lenk (Ed.), Handlungstheorien – interdisziplinär. Band III, Teilband 2, im Druck.

Herzlich, C. *Health and illness. A social psychological analysis.* London: Academic Press, 1973.

Hewstone, M., & Jaspars, J. Intergroup relations and attribution processes. In H. Tajfel (Ed.), *Social identity and intergroup behaviour.* Cambridge: Cambridge University Press, and Paris: Maison des Sciences de l'Homme, 1982.

Hilke, R. Wie aggressiv sind Versuchspersonen wirklich? Einige Überlegungen zur "klassischen" experimentellen Versuchsanordnung aus Anlaß eines nicht alltäglichen empirischen Befundes. *Zeitschrift für Sozialpsychologie,* 1977, *8,* 137–155.

Kaufmann, H. Definitions and methodology in the study of aggression. *Psychological Bulletin,* 1965, *64,* 351–364.

Kaufmann, H. *Aggression and altruism: A psychological analysis.* New York: Holt, Rinehart & Winston, 1970.

Krauth, J., & Lienert, G. A. KFA. Die Konfigurationsfrequenzanalyse. Freiburg: Alber, 1973.

Lagerspetz, K. M., & Westman, M. Moral approval of aggressive acts: A preliminary investigation. *Aggressive Behavior,* 1980, *6,* 119–130.

Lalljee, M. Attribution theory and the analysis of explanations. In C. Antaki (Ed.), *The psychology of ordinary explanations of social behaviour.* London: Academic Press, 1981.

Lienert, G. A., & Krauth, J. Configural frequency analysis as a statistical tool for defining types. *Educational and Psychological Measurement,* 1975, *35,* 231–238.

Linneweber, V. *Aggressive Interaktionen. Eine feld- und verhaltensspezifische Taxonomie von Interaktionssituationen in Schulen.* Unpublished doctoral dissertation, Universität Münster, 1981.

Linneweber, V., Mummendey, A., Bornewasser, M., & Löschper, G. Classification of situations specific to field and behavior: The context of aggressive interactions in schools. *European Journal of Social Psychology,* 1984, *14,* in press.

Löschper, G. *Definitionskriterien aggressiver Interaktionen. Normabweichung, Intention und Schaden als Einflußfaktoren auf die Definition von Verhaltensweisen als aggressiv.* Unpublished doctoral dissertation, Universität Münster, 1981.

Löschper, G., Mummendey, A., Linneweber, V., & Bornewasser, M. The judgement of behavior as being aggressive and sanctionable. *European Journal of Social Psychology*, 1984, *14*, in press.

Moos, R. H. *Evaluating treatment environments. A social-ecological approach.* New York: Wiley, 1974.

Moscovici, S. On social representations. In J. P. Forgas (Ed.), *Social cognition. Perspectives on everyday understanding.* London: Academic Press, 1981.

Mummendey, A. Zum Nutzen des Aggressionsbegriffs für die psychologische Aggressionsforschung. In R. Hilke & W. Kempf (Eds.), *Aggression. Naturwissenschaftliche und kulturwissenschaftliche Perspektiven der Aggressionsforschung.* Bern: Huber, 1982.

Mummendey, A., Bornewasser, M., Löschper, G., & Linneweber, V. Aggressiv sind immer die anderen. Plädoyer für eine sozialpsychologische Perspektive in der Aggressionsforschung. *Zeitschrift für Sozialpsychologie*, 1982, *13*, 177–193.

Mummendey, A., Linneweber, V., & Löschper, G. Actor or victim of aggression: Divergent perspectives – divergent evaluations. *European Journal of Social Psychology*, 1984, *14*, in press.

Mummendey, A., Löschper, G., Linneweber, V., & Bornewasser, M. Social-consensual conceptions concering the progress of aggressive interactions in school. *European Journal of Social Psychology*, 1984, *14*, in press.

Nagel, T. Ruthlessness in Public life. In S. Hampshire (Ed.,) *Public and private morality.* Cambridge: Cambridge University Press, 1978.

Nesdale, A. R., Rule, B. G., & McAra, M. Moral judgements of aggression: Personal and situational determinants. *European Journal of Social Psychology*, 1975, *5*, 339–349.

Newcomb, T. M., Turner, R. H., & Converse, P. E. *Social psychology: The study of human interactions.* New York: Holt, Rinehart & Winston, 1965.

Newtson, D. Foundations of attribution: The perceptions of ongoing behavior. In J. H. Harvey, W. Ickes, & R. F. Kidd (Eds.), *New directions in attribution research.* Vol. 1. New York: Erlbaum, 1976.

Nickel, T. W. The attribution of intention as a critical factor in the relation between frustration and aggression. *Journal of Personality*, 1974, *42*, 482–492.

O'Neal, E. C., & McDonald, P. J. The environmental psychology of aggression. In R. G. Geen, & E. C. O'Neal (Eds.), *Perspectives on aggression.* New York: Academic Press, 1976.

Pettit, P. Rational man theory. In C. Hookway & P. Pettit (Eds.), *Action and interpretation.* Cambridge: Cambridge University Press, 1978.

Pettit, P. On actions and explanations. In C. Antaki (Ed.), *The psychology of ordinary explanations of social behaviour.* London: Academic Press, 1981.

Price, R. H. The taxonomic classification of behaviors and situations and the problems of behavior-environment congruence. *Human Relations*, 1974, *27*, 567–585.

Price, R. H., & Bouffard, D. L. Behavioral appropriateness and situational constraint as dimensions of social behavior. *Journal of Personality and Social Psychology*, 1974, *30*, 579–586.

Price, R. H., & Moos, R. H. Toward a taxonomy of inpatient treatment environments. *Journal of Abnormal Psychology*, 1975, *84*, 181–188.

Rule, B. G., & Nesdale, A. R. Moral judgement of aggressive behavior. In R. G. Geen, & E. C. O'Neal (Eds.), *Perspectives on aggression.* New York: Academic Press, 1976.

Shaw, M. E., & Reitan, H. T. Attribution of responsibility as a basis for sanctioning behaviour. *British Journal of Social and Clinical Psychology*, 1969, *8*, 217–226.

Stapleton, R. E., Joseph, J. M., & Tedeschi, J. T. Person perception and the study of aggression. *Journal of Social Psychology*, 1978, *105*, 277–289.

Stokols, D. *Perspectives on environment and behavior. Theory, research and applications.* New York: Plenum Press, 1977.

Stokols, D. Group × place transactions: Some neglected issues in psychological research on settings. In D. Magnusson (Ed.), *Toward a psychology of situations: An interactional perspective.* Hillsdale, New Jersey: Erlbaum, 1980.

Stroebe, W. The level of social psychological analysis: A plea for a more social social psychology. In L. H. Strickland (Ed.), *Soviet and western perspectives in social psychology.* Oxford: Pergamon Press, 1979.

Tajfel, H. Intergroup relations, social myths and social justice in social psychology. In H. Tajfel

(Ed.), *The social dimension: European developments in social psychology*. Cambridge: Cambridge University Press (in press)

Tajfel, H. Experiments in a vacuum. In J. Israel & H. Tajfel (Eds.), *The context of social psychology. A critical assessment*. London: Academic Press, 1972.

Tajfel, H. *Human groups and social categories*. Cambridge: Cambridge University Press, 1981.

Tajfel, H. Psychological conceptions of equity: The present and the future. In P. Fraisse (Ed.), *Psychologie de demain*. Paris: Presses Universitaires de France, 1982.

Tajfel, H., & Turner, J. An integrative theory of intergroup conflict. In W. G. Austin & S. Worchel (Eds.), *The social psychology of intergroup relations*. Monterey, Calif.: Brooks/Cole, 1979.

Tedeschi, J. T., & Lindskold, S. *Social psychology. Interdependence, interaction and influence*. New York: Wiley & Sons, 1976.

Tedeschi, J. T., & Riess, M. Verbal strategies in impression management. In C. Antaki (Ed.), *The psychology of ordinary explanations of social behaviour*. London: Academic Press. 1981.

Tedeschi, J. T., Smith, R. B., & Brown, R. C. A reinterpretation of research on aggression. *Psychological Bulletin*, 1974, *81*, 540–562.

Weinstein, E. A. The development of interpersonal competence. In D. Goslin (Ed.), *Handbook of socialization theory and research*. New York: Rand McNally, 1969.

Werbik, H. Das Problem der Definition "aggressiver" Verhaltensweisen. *Zeitschrift für Sozialpsychologie*, 1971, *2*, 233–247.

Werbik, H., & Munzert, R. Kann Aggression handlungstheoretisch erklärt werden? Psychologische Rundschau, 1978, *29*, 195–208.

Zillmann, D. *Hostility and aggression*. Hillsdale, New Jersey: Erlbaum, 1979.

Zuckerman, M. Attribution of success and failure revisited, or: The motivational bias is alive and well in attribution theory. *Journal of Personality*, 1979, *47*, 245–287.

Acknowledgment. The studies reported within this contribution are part of a research project supported by a grant from *Deutsche Forschungsgemeinschaft* (Az. Mu 551).

The authors are deeply indebted to Dipl.-Soz. Detlef Fichtenhofer for providing the major part of the statistical data analysis and to Sabine Otten and Jörg-Dietrich Meyberg for performing a great deal of the empirical studies.

Chapter 6

Patterns of Aggressive Social Interaction

Richard B. Felson

This paper has two purposes. First, I discuss three interactionist approaches to aggression that can be found in the literature. Most attention is given to an approach that interprets aggression as punishment for perceived wrongdoing, since that approach has never been fully explained. I argue that each approach is useful but limited in its ability to account for how aggressive encounters develop. Then, using some of the ideas developed in the theoretical section, I analyze self-reports of aggressive encounters of different levels of severity. An attempt is made to describe in a theoretically informed way what occurs in interactions that culminate in an aggressive attack.

Aggression is defined here as an act in which a person attempts or threatens to harm another person, regardless of the ultimate goal of the act. Thus, it is assumed that aggression (including "angry aggression") is a means to an end, rather than an end in itself, and that persons use harm for a variety of purposes. For example, aggression may be used to save face, to teach someone a lesson, to defend oneself, or even to ultimately benefit the target. The definition includes behavior that is sometimes described as punishment, since punishment involves an attempt to harm another person. Harm-doing is often labeled as punishment, rather than aggression, when it is perceived as legitimate, but value judgments should be avoided in defining a phenomenon. And, as we shall see below, interpreting aggression and punishment as the same type of behavior (and showing the legitimacy of many aggressive acts) has significant implications for explaining how many aggressive interactions begin.

Theoretical Approaches

Aggression as Impression Management

The first approach is based on symbolic interactionism and a derivative of that approach, impression management theory. It focuses on the role of the self in aggressive situations, interpreting aggression as face-saving behavior that occurs when one perceives oneself as having been intentionally attacked. For example, in a previous paper (Felson, 1978) I suggested that when persons have been cast into negative situational identities (by insults, for example), they retaliate in order to resist those identities (see also, e.g., Hepburn, 1973; Luckenbill, 1977; Athens, 1980). By casting the original attacker into a negative identity, the identity implied by the original attack is negated, and the person's honor or public image is restored. Retaliation is particularly likely if third parties are present, since face-saving concerns are exaggerated. However, if third parties mediate, the conflict is likely to be less serious, since both sides can back down without losing face (Rubin, 1980; Felson, Ribner & Siegel, in press).

This approach is limited in two important respects. First, it does not explain aggression that is used strategically to produce tangible rewards and to avoid costs. For example, force is used in robberies primarily for strategic purposes (Luckenbill, 1980). Second, the approach is limited in that it explains retaliatory aggression, but not the initial attack. Face-saving and maintaining honor are only relevant once persons think they have been attacked. With some notable exceptions (e.g., violence by youthful gangs against outsiders), unprovoked attacks are viewed negatively. In fact, the approach suggests that persons avoid unprovoked attacks on others because of rules of deference (Goffman, 1956), which protect identities in interaction, and because of the embarrassment that ensues when someone's identity has been threatened. In some instances the original attack may be inadvertent; all that is necessary is that persons believe that they have been attacked and an aggressive conflict is likely to begin. However, it seems unlikely that in most instances the initial attack is inadvertent.

Aggression as Coercive Power

A second interactionist theory of aggression has been suggested by Tedeschi and his associates (e.g., Tedeschi, 1970; Tedeschi, Smith & Brown, 1974; Tedeschi, Gaes & Rivera, 1977). Tedeschi's theory of coercive power suggests that harmdoing – the concept of "aggression" is avoided because of its value connotations – is simply one technique used to influence people when other methods fail. Harm and the threat of harm ("punishment and threats"), whether provoked or not, are used to influence others to provide rewards. The decision to attack an antagonist depends on a calculation of the rewards and costs that would result from such an attack.

Unlike the impression management approach, Tedeschi's scheme can explain initial attacks as well as retaliation. Further, it handles strategic factors and can explain why the target of an attack might choose to submit, rather than retaliate. It is also limited, however, in two important ways. First, harmful acts that involve face-saving are not necessarily coercive, i.e., they are not necessarily used to force others to comply. For example, a young male who fights to avoid losing face in front of his girlfriend is not attempting to *force* anyone to change their behavior, since one cannot usually coerce an audience to give its approval. Rather, he is behaving in a way that he thinks will bring him approval from someone whose opinion he values. Only if he fights to display his power to his adversary or other would-be adversaries, or to maintain the credibility of his threats, could it be said that he was engaging in coercive power. Thus some, but not all, impression managment is coercive.

Second, not all harmful acts are designed to influence others. For example, harm that is used to maintain equity or promote justice is not an attempt at social influence. In other words, some harm-doing is perceived as legitimate or just when someone has done something that deserves punishment. This will be discussed in greater detail below.

Aggression as Punishment

Aggression can also be conceptualized as punishment – either actual or threatened (Felson, 1981; 1982).[1] Such an approach implies that the harmful act is in response to some offense or misbehavior. It also suggests the possibility that harm-doing can be legitimate when it is identified as punishment for a wrongful act. Aggresssion is more easily understood if it can be shown to be perceived as legitimate or socially desirable in certain instances.[2]

This approach can be applied to both the legal system and to more informal acts of punishment. In the case of the legal system, the state specifies penalties for various criminal offenses in the form of fines or prison sentences and aggression is institutionalized. In the case of noncriminal offenses, the punishment may range from reproaches for wrongful behaviors to verbal and physical attacks focused directly on the offender. Persons who observe an offense are usually entitled to at least reproach an offender, and persons in special roles, e.g., parents and teachers, have the specific authority to punish. In fact, for more serious offenses, observers may be expected to punish the offender.

Black (1983) has made a similar argument in his interpretation of some types of crime as social control:

"Far from being an intentional violation of a prohibition, much crime is moralistic and involves the pursuit of justice. It is a mode of conflict management, possibly a form of punishment, even capital punishment. Viewed in relation to

---

1 Tedeschi also discusses threats and punishments, but coercive power appears to be his central focus. He recognizes the limitations of the coercive power approach and is attempting to develop a more comprehensive theory (personal communication).
2 Aggression is often elicited in experimental situations by legitimating it as punishment for wrong-doing (e.g., the teacher-learner paradigm).

law, it is self-help. To the degree that it defines or responds to the conduct of someone else – the victim – as deviant, crime is social control." (Black, 1983, p. 34)

For Black, self-help is an expression of a grievance by unilateral aggression that was more common before the advent of law and formal legal systems, but that continues to exist.

The rationale or explanation for legal punishment and informal punishment are the same. Using the language of legal theory, two explanations are generally given: deterrence and retribution. Deterrence refers to the use of punishment to deter future offenses and directly corresponds to the notion of coercive power discussed by Tedeschi. It includes specific deterrence (the threat of punishment to the offender for future violations), general deterrence (the threat to third parties who might consider engaging in such violations), and incapacitation (rendering the offender unable to commit another offense). From the perspective of the group, then, aggression is an important mechanism of social control. Groups encourage harm-doing in certain instances to discourage and control deviance.

Retribution refers to the desire of persons to see misbehavior punished, even when deterrence is not an issue. It is sometimes described in terms of a norm of justice or distributive justice or equity. The norm can be used to explain why people want to harm, or see others harm, persons who are guilty of some wrongful act, even if they are not the victim of that act. Everyday observation suggests that persons can become quite angry about an injustice, i. e., an offense that goes unpunished, even when they are personally unaffected. Further, the norm explains a basic fact about punishment: the severity of punishment is strongly related to the severity of the offense (Durkheim, 1947).[3]

Unfortunately, this interactionist approach to aggression is also limited. First, it may be stretching it a bit to consider harm-doing during robberies and rapes as punishment. True, violence is used primarily during these crimes in response to, or in anticipation of, noncompliance and thus is used to deter resistance. However, this noncompliance may or may not be viewed by the offender as an offense. The question is whether the robber or rapist views his own behavior or his target's behavior as legitimate. Second, the approach does not account for all instances in which aggression is used to save face. When persons are victims of a personal attack, as opposed to some other type of offense or as opposed to an offense against some other party, they are particularly likely to harm the perpetrator. In other words, retaliation is likely because the target's identity is involved. And, in contrast to retribution, the victim of an attack on identity seeks victory in the aggressive encounter, not a fair exchange where the punishment and the offense are proportionate. Both face-saving and retribution may play a role in retaliation for a personal attack, since this behavior is both an attack on identity and a wrongful act. Targets want to retaliate in order to save face and are justifi-

---

3 Durkheim (1947) explained punishment by reference to a "collective conscience" rather than a norm of retribution and he rejected the deterrence argument as an explanation of punishment. Goode (1973) argues that most punishment at the interpersonal level is not consciously intended as a deterrence strategy.

ed in doing so because the aggressor has violated a norm and deserves punishment. Thus the norm of justice serves the motive for revenge.

The punishment approach suggests that a social control process occurs in the early stages of an aggressive encounter. This process culminates in a punishment response in the form of a reproach or an insult. I now describe the actions that are likely to precede the punishment.

1. Type of Offense. There are two types of offenses that may result in punishment: violations of norms and violations of orders. In the case of norm violations, offenders are punished for something they have already done. In the case of violation of orders, punishment occurs when a person will not comply with a request or command. Similarly, Schelling (1960; 1966) distinguishes compliance from deterrence, the former referring to actions designed to get persons do what they would otherwise not do, rather than to cease what they are already doing.

Violations of norms and orders are similar conceptually. As Homans (1950) suggests: "Orders are not different in kind from norms. Both norms and orders ... specify what the behavior of the members of a certain group ought to be rather than what it really is. The only difference between the two is that norms apply to the maintenance of established behavior, orders to future changes in behavior ... "(p.416). In addition, violations of norms may result in orders to cease the offensive behavior. Continuation of that behavior would then involve noncompliance with both norms and orders.

Informal punishment for norm violations has been observed in a number of field studies. For example, in his study of the bank wiring observation room, Homans (1950) found that ridicule and "binging" – punching in the arm – were used against rate-busters. Miller, Geertz, and Cutter (1961) found that most of the aggressive acts of the boys' group they observed were directed against group members for "failure to contribute toward group ends, disruption of concerted activities, or failure to maintain expected relations of reciprocity and equality" (p.296). Punishment for noncompliance with orders has been observed in studies of criminal violence. The frequent occurence of requests, commands, threats, and noncompliance in homicides and assaults (not involving robbery and rape), and the infrequency of compliant actions, have been reported by Luckenbill (1977) and Felson and Steadman (1983). For example, Felson and Steadman found that homicides and assaults began with verbal conflict in which identities were attacked and attempts to influence the victim failed. These were followed by threats and evasive actions and, finally, physical attack.

2. Accounts. Punishment may not occur immediately after the audience perceives a violation. According to the literature on accounts, the audience may respond with a challenge or query as to why the offender has engaged in such an action. Either in response to these challenges or in anticipation of them, offenders may provide accounts. An account is an excuse or justification for some potentially deviant act (Scott & Lyman, 1968). By giving accounts, offenders attempt to align themselves with the normative order and divorce themselves from their actions (Stokes & Hewitt, 1976). When accounts are successfull, offenders can avoid sanctions from the audience, or at least reduce their severity. Thus, Felson (1982) found that if either antagonist in an aggressive situation provided an account of his actions, the verbal conflict was less likely to become physically violent.

Accounts may occur in three possible positions during a social control sequence; (1) before the rule violation (see Hewitt & Stokes, 1975, on "disclaimers"), (2) immediately following rule violations, or (3) after the reproach. In the first two instances the actor anticipates reproaches and other negative reactions from the audience and attempts to avoid them. One suspects that many aggressive interactions are avoided in this way. If the actor fails to anticipate the negative reaction and is openly reproached by the other, an aggressive interaction is likely, since that reproach may be perceived as an attack. Thus, in most aggressive interactions we would expect that persons have failed to give accounts until they have been reproached. As a result, one would expect accounts to occur following reproaches, rather than before or immediately after rule violations.

A Comparison of the Approaches

I have reviewed three interactionist approaches to aggression. The first emphasizes identities and face-saving, the second emphasizes coercion to produce compliance, and the third emphasizes punishment for perceived wrongdoing. All three are limited in the types of behaviors they can explain. Impression management theory is limited because it cannot account for the first attack or for strategic factors. For example, it cannot explain why persons sometimes do not retaliate against a more powerful opponent, even if it means a loss of face. Tedeschi's theory of coercive power is limited because it cannot account for impression management behavior (unless it involves a display of power) or for legitimate retribution. Finally, the interpretation of aggression as punishment is limited because it cannot handle face-saving behavior or coercion where the targets' noncompliance is viewed as legitimate. Unfortunately, there is no single scheme that can account for every instance of aggression, and therefore a more parsimonious theory is unavailable. However, integration may be possible, since the three schemes are similar in a number of ways. First, all three are interactionist theories in that they emphasize situational factors and the importance of the interaction preceding the attack. Second, all three interpret aggression as social influence behavior, although the punishment approach emphasizes the normative aspect of this behavior as well. Finally, the coercive power approach and the punishment approach both emphasize the role of aggression in producing compliance.

Perhaps an integration would be possible if either the pursuit of favorable identities or the desire for tangible rewards is identified as more basic. From a symbolic interactionist point of view, for example, persons may pursue rewards and avoid costs (through force and other means), and follow the norm of retribution (and other norms), in their pursuit of favorable identities. On the other hand, a reinforcement position would suggest that pursuit of identities and conformity to norms are attributable to the desire for tangible benefits. One's choice in these matters is largely a function of one's philosophical predilections and I will not pursue the argument here. Instead, I borrow from all three approaches (although not equally) to suggest the following scenario. Aggressive encounters generally begin when persons violate norms or orders. The audience responds to these violations with an attack or a threat to attack, either because they wish to produce compliance (by the target or third parties) or because they believe wrongdoing deserves to be punished. This initial attack can be described, then, as social control behavior. Once an attack of any kind occurs, identities and face-saving become involved and the likelihood of further attack is increased. However, the target of an attack may submit if he believes the costs of retaliation are too high.[4]

A Description of Aggressive Incidents

Introduction

In the empirical portion of this paper, I attempt to describe what occurs in episodes culminating in intentional attacks. I shall use the theoretical scheme just described to analyze these episodes, but by no means do I view these analyses as a test of this perspective. Incidents of four different levels of severity are examined: incidents in which the respondent was angry, but did nothing about it; incidents involving verbal aggression only; incidents involving physical violence, but no weapons; and incidents involving weapons.

Three types of samples are studied: the general population, ex-mental patients, and ex-offenders. One would expect that the differences among these populations are extreme in terms of a variety of individual characteristics. Therefore, findings of similar relationships between variables for all three populations would provide evidence for the generality of the processes involved.

I begin by examining the frequencies of different actions that the respondents attribute to their antagonists and to themselves in these incidents. The discussion above and our earlier work with crime data suggest that the following types of actions are likely to occur during an aggressive encounter: orders, noncompliance, rule violations, reproaches, accounts, insults, threats, submission, and physical attacks. Log linear analysis is used to examine whether the frequencies of these actions are associated with the severity of the incident, with controls for the type of sample and the sex of the actor. I also discuss sex differences in these

---

4 An interactionist approach also has implications for the types of individual variables that are likely to be associated with aggression. The following variables are likely to be important: the propensity to break rules (and thus become a target of punishment); the ability and willingness to provide accounts, particularly before one has been reproached; the tendency to accept or believe the accounts of others (i.e., trust); the ability to use other techniques to influence others effectively; an aversion to creating "scenes" or embarrassment; harshness in the evaluation of various offenses; judgments of appropriate punishment, in particular, the willingness to use corporal punishment; the importance of maintaining one's honor; and the tendency to perceive reproaches or other behavior as an attack on the self.

behaviors. One expects that males are more aggressive than females, since most studies find evidence of sex differences in aggressive behavior (see Frodi, Macaulay & Thome, 1977, for a review). Second, I examine the ordering or position of actions during these incidents to determine whether social control behavior occurs early and verbal and physical attacks occur later in the incident. Third, I attempt to determine the position of accounts in the social control sequence by examining the sequencing of accounts, rule violations, and reproaches. I expect that accounts will tend to occur after reproaches because persons do not anticipate (sometimes by choice) negative reactions from the audience to their rule violations. Finally, I examine whether the agent of control or the target is the first to attack. If the initial attack is punishment for some violation, then the social control agent should be the first to attack. However, in some instances it may be that the control agent responds to the offense with a reproach and that the target is the first to attack, in retaliation for the reproach.

Samples

The analyses are based on interviews of persons aged 18–65 from the general population ($N = 245$), ex-mental patients ($N = 148$), and ex-criminal offenders ($N = 141$) in Albany County, New York. A representative sample from the general population was obtained through a multistage process in which street names were randomly chosen from each of 35 census tracts based on the percentage of population in that tract. A sample of dwellings on each street was then randomly selected, and a determination was made as to whether a male or female was to be interviewed, in order to achieve an equal sex distribution.

The sample of ex-mental patients included persons who had been living in the community for at least 6 months in the year preceding the interview. Respondents were contacted either through the mail or through visits to social clubs in the area for ex-mental patients.

The ex-offenders included parolees and local offenders who were contacted by mail. In addition, contact was made through a community day program for women who had been released from the state prison or from a local jail. All respondents had been living in the community for at least 6 months in the year preceding the interview.

Measurement

Respondents were asked to describe in detail four incidents of varying severity. They were told that this part of the interview would be recorded so that no information would be lost, but that their responses were confidential and that the tapes would be erased in a few days. Then they were asked to "recall the last dispute that you can remember clearly, that you were involved in, where a gun or knife was drawn or used." Later on they were asked a similar question about a dispute where there was "slapping or hitting with a fist, but no gun or other weapon was involved." The third incident they were asked to describe involved a "bad argument with someone which involved screaming, shouting, or name-calling, but not slapping or hitting – in other words, just a dispute where there was a bad argument, but nothing physical." Finally, they were asked to recall a dispute in which they were "really angry at another person, but said nothing about it. In other words, we mean you kept your anger inside."

Each description was coded as a sequence of unit-actions with an actor identified for each action. Three types of actors were identified: the respondent, the main antagonist, and third parties, i.e., anyone else present during the incident who engaged in some action. The actions were classified according to a scheme developed in earlier work on homicide and assault (Felson & Steadman, 1983). After coding the actions in detail, we classified them into ten general categories:

1. *Physical attacks,* including physical violations, pursuing for physical attack, and drawing and struggling for a weapon.
2. *Insults* or direct attacks on identity, including instances of yelling.
3. *Threats,* including challenges and dares and nonverbal threatening gestures. Here the actor indicates that harm is forthcoming; some but not all of these are contingent threats, i.e., communications of impending harm unless the target complies.
4. *Rule violations,* including annoying behavior, failure to discharge an obligation, ignoring, causing another's loss inadvertently, boasting, inappropriate demeanor, infidelity, taking someone's property, or violating that property.
5. *Reproaches,* including accusals, complaints, protests, commands to cease some offensive action or to leave, chastisement, and asking for accounts or redress. These are social control actions that focus on the behavior of their target, although they have implications for identities.
6. *Accounts,* or explanations of conduct.
7. *Submission,* including apologies, compliance, crying, pleas not to attack, and fleeing.
8. *Orders,* including requests and commands, except commands to cease offensive actions, since the latter respond to previous wrongful action. These are compellent or persuasive actions, i.e., actions designed to produce compliance. However, this is not an ideal label, since orders imply that compliance is obligatory, while at least some (but not all) requests imply that compliance is optional. That is, in some contexts, actions that appear on the surface to be requests are actually commands.
9. *Noncompliance,* including refusal to comply and doing nothing when the antagonist has called for compliance.
10. *Mediation,* or actions that attempt to reconcile the opposing parties.

These were coded according to the ostensible nature of the action; no attempt was made to determine the underlying purpose or effect of the action. Further, only specific actions, and not emotions, were coded. Each tape-recorded event was independently coded by the interviewer and the project director. The coders agreed on the actions and their order approximately 78% of the time. The coders met afterward to resolve their differences and to agree to a final code.

While I recognize the limitations of self-report data, I would argue that these data are adequate for our purposes. Although respondents may forget some actions that occurred during the conflict and although they may attribute more negative actions to the antagonist and acknowledge fewer themselves, these biases should not affect the major conclusions. It is doubtful that such reporting biases can readily explain the relationships observed among various actions, sex, and severity. It is also unlikely that the order of events or the relationship between committing the first attack and engaging in rule violations or orders is affected by such biases. It is probable, however, that the patterns observed would have been more clear-cut if there were no measurement error. Moreover, there is reason to believe that the absolute frequencies of actions are underestimated due to memory loss. In sum, while self-report data are imperfect, they are the best available for describing what occurs in conflicts of these types among adults. Thus, I believe that an examination of individual findings, with an eye to alternative artifactual explanations, is more appropriate than a total rejection of the method.

Action Frequencies by Sex and Type of Incident

The actions of the respondent and antagonist are presented for each type of incident for the general population in Table 6.1.[5] Since these absolute frequencies are probably underestimated, it is better to consider the relative frequencies of different actions to determine their role. According to these results, rule violations, reproaches, accounts, orders, and noncompliances occur frequently, suggesting that the social control process is playing an important role in many of these incidents.

Table 6.1. Percentages of incidents where respondents and antagonists engaged in various actions (general population only)

Actions	Respondents situations				Antagonists			
	Only anger	Verbal	Hitting	Weapon	Only anger	Verbal	Hitting	Weapon
Rule violation	6.0	13.8	18.2	14.0	43.5	26.1	25.2	20.9
Orders	19.0	23.6	14.7	27.9	13.6	20.2	9.8	30.2
Reproaches	26.1	59.1	44.1	37.2	23.9	58.6	32.2	41.9
Noncompliance	2.3	11.8	16.1	18.0	6.9	13.8	13.3	9.3
Accounts	22.3	44.3	16.8	25.6	15.2	28.6	14.0	7.0
Insults	3.8	23.2	22.4	18.6	11.4	33.5	39.9	46.5
Threats	4.3	14.3	17.5	23.3	4.9	10.3	18.9	30.2
Physical attack	–	–	78.3	41.9	–	–	76.2	86.0
Submission	5.4	9.9	14.0	27.9	10.9	15.3	17.5	18.6

5 Note that respondents attribute more aggressive actions and rule violations and fewer accounts to the antagonists than they acknowledge themselves. There are two explanations for these differences. First, respondents may be biased either in their perception of the incident as

Log linear analyses were performed to determine whether either the respondent's or the antagonist's actions varied by population, sex, or type of incident. The data set was rearranged so that the incident was the unit of analysis. Each model involved five variables: Type or severity of incident (S) × sample or population (P) × sex of respondent (R) × sex of the antagonist (A) × action (V). The first four variables are the same in all models; the fifth (V) involves one of the nine actions of either the respondent or antagonist, producing a total of $9 \times 2 = 18$ models.[6] Most analyses are based on a 4 (angry, but did nothing; argument; hitting; weapon) × 3 (general population, offenders, ex-patients) × 2 (male, female) × 2 (male, female) × 2 (no, yes) table with 96 cells and an N of

Table 6.2. Final models involving respondent's and antagonist's action

Actions	Respondent's action			Antagonist's action		
	Model	LRχ^2	p-level	Model	LRχ^2	p-level
Rule violations	PRA, SRA, PS, SV, RV	47.8	0.89	PRA, SRA, SAV, PS	42.3	0.94
Request/orders	PRA, SRA, PS, V	49.0	0.93	PRA, SRA, PS, SV, RV	39.9	0.98
Reproaches	PRA, SRA, PS, PV, SV, RV	58.4	0.50	PRA, SRA, RAV, PS, SV	44.7	0.92
Noncompliance	PRA, SRA, PS, SV	49.0	0.88	PRA, SRA, PS, V	49.5	0.92
Accounts	PRA, SRA, PS, PV, SV	49.7	0.83	PRA, SRA, PS, PV, SV	46.7	0.89
Insults	PRA, SRA, PS, PV, SV	46.0	0.91	PRAV, SRA, SRV, PSV	41.2	0.51
Threats	PRA, SRA, PS, PV, SV, AV	48.5	0.83	PRA, SRA, SAV, PS, PV	51.8	0.64
Physical attack	PRA, PSV, SRV, SA	29.9	0.15	PRA, RAV, PS, SA	21.4	0.81
Submission	PRA, SRA, PS, SV	55.3	0.68	PRA, SRA, PSV	39.3	0.93

Note. P, sample; R, respondent's sex; A, antagonist's sex; S, severity of incident; V, the action in the left-hand column, committed by either respondent or antagonist, as described by the caption.

---

it occurred, in their recall of it, or in their willingness to report what they remember to the interviewer. Thus, they may attribute more negatively valued actions (aggression and rule violations) to the antagonist and claim more positively valued actions (accounts) to themselves. Note that respondents attribute similar frequencies of neutrally valued actions to themselves and the antagonist. Second, antagonists in these incidents may actually be more aggressive than the respondents. This is likely because the sample of antagonists was chosen because of their participation in aggressive incidents, while the respondents were not. This is supported by the finding that differences in aggression between respondents and antagonist are most notable in the weapons incidents, and these incidents are more likely to sample aggressive antagonists. We suspect that both of these processes are operating. In subsequent analyses we shall have to assume that the motivational bias is not altering our results in a systematic way.

6 While this is a large sample, it is not large enough to incorporate more than one action in a single model. Also, since these analyses focus on the actions of respondents and antagonists, mediation, which is performed primarily by third parties, is not included.

1326. The large N is due to the fact that most respondents described multiple incidents. Since the analysis of physical attacks by the respondent and antagonist are only performed on the incidents involving physical violence, these analyses are based on a $2 \times 3 \times 2 \times 2 \times 2$ table with 48 cells and an N of 551. A value of 0.5 is added to each cell because a few of the cells had zero entries (Goodman, 1970).

The final log linear models are presented in Table 6.2. While the fit of the model for respondent's physical attacks is only adequate, the rest of the fits are excellent. Each model includes a saturated term for the independent variables (PRA) as is customary. In the models involving all four types of incidents, there is a sex of respondent $\times$ sex of antagonist $\times$ severity interaction (SRA: $LR\chi^2 = 29.4$ (3); $p < 0000$).[7] This is due to the fact that conflicts are more likely to involve physical violence when males are in conflicts with other males. There is also a sample $\times$ severity interaction, reflecting the fact that incidents involving ex-offenders tend to be more severe (PS: $LR\chi^2 = 40.7$ (6); $p < .0000$). We now describe, in turn, the interactions involving different actions and the severity of the incident (S), and the sex of the respondent (R) and antagonist (A).

1. Severity of the Incident. The antagonist was most likely to engage in a rule violation in incidents where the respondent was angry, but did nothing.[8] This suggests that rule violations by the antagonist often do not elicit an overt response from the respondent. If the antagonist engages in a reproach, on the other hand, the respondent is unlikely to withhold his anger (SV: $LR\chi^2$ (3) = 93.9; $p < .0000$). One suspects that reproaches are more likely to be perceived as attacks on identity than are rule violations, and thus retaliation is more likely.

Insults and threats also appear to lead to an escalation of the incident. As the severity of the incident increases, the frequency of threats from the respondent increases (SV: $LR\chi^2$ (3) = 50.4; $p < .0000$). Severity is also strongly associated with antagonist's threats, particularly for females (SAV: $LR\chi^2$ (3) = 10.3; $p = 0.02$). Insults from the antagonist are positively related to severity, but the effect is masked by some marginally significant ($p = 0.05$ and $p = 0.06$) interactions involving sample (SVP) and respondent's sex (SVR). The omission of the severity X insult from the antagonist (SV) term from a full two-order model is highly significant ($LR\chi^2$ (3) = 57.0; $p < .0000$). The pattern for respondent's insults is not as clear-cut, since these insults are most frequent in verbal disputes and least frequent in incidents involving unexpressed anger (SV: $LR\chi^2$ (3) = 66.7; $p < 0.0000$).

Accounts from either party appear to reduce the probability of physical violence. Accounts from the respondent (SV: $LR\chi^2$ (3) = 70.2; $p < 0.0000$) and from

---

7 Unless otherwise noted, significance tests are based on the significance of differences in likelihood-ratio chi squares when the term is omitted from the final model.

8 This effect is masked by an SVA term which results because males engage in more rule violations than females in arguments and fewer rule violations than females in hitting incidents. The relationship between unexpressed anger and rule violations is unaffected by sex. The omission of the SV term from a full two-order model significantly decreases the adequacy of the fit ($LR\chi^2$ (3) = 29.7, $p < 0.0000$).

the antagonist (SV: $LR\chi^2$ (3) $= 61.1; p < 0.0000$) were more frequent in incidents involving verbal disputes and unexpressed anger than in incidents involving physical violence.

Finally, the greater the severity of the incident, the greater the likelihood of a submissive action. Thus, the number of submissive actions by the respondent is positively associated with the severity of the incident (SV: $LR\chi^2$ (3) $= 30.0$; $p < 0.0000$). A similar pattern is oberved for antagonist's submission, although it appears to vary slightly across populations (PSV: $LR\chi^2$ (6) $= 13.3; p = .04$). The omission of the SV term from a full two-order model is highly significant ($LR\chi^2$ (3) $= 16.8, p = 0.0008$).

In sum, the evidence suggests that rule violations are a source of many of these conflicts, but that rule violations can be ignored and open conflict avoided. Once one party reproaches or insults the other a verbal conflict is likely, since the other party is likely to retaliate. If the initial rule violator gives an account for his actions, the conflict is unlikely to become more serious, i.e., become physical. But if threats are made, physical violence and escalation (more generally) are likely. Finally, as escalation and the potential costs of the conflict increase, so does submission.[9]

2. Sex effects. In general there are no sex differences in verbal aggression (insults and threats), but there are strong differences in physical aggression. Male respondents are more likely to engage in physical attacks in hitting incidents (SRV: $LR\chi^2$ (1) $= 8.4; p = 0.004$), and male antagonists are more likely than female antagonists to engage in physical attack when the respondent is male (RAV: $LR\chi^2$ (1) $= 9.8; p = 0.002$). This is consistent with the interaction noted earlier among severity, sex of respondent, and sex of antagonist.

There are also sex differences in social control behavior. Female respondents are more likely to engage in reproaches (RV: $LR\chi^2$ (1) $= 8.6; p = 0.003$), and female antagonists are more likely to reproach males than male antagonists (RAV: $LR\chi^2$ (1) $= 8.6; p = 0.003$). Also, while male respondents are slightly more likely to engage in rule violations than female respondents (RV: $LR\chi^2$ (1) $= 3.7$; $p = 0.056$), the pattern for antagonist's rule violations is unclear.

---

9 Ex-offenders engage in the most verbally aggressive actions of the three populations in that they are most likely to engage in threats (PV: $LR\chi^2$ (2) $= 13.6; p = 0.001$) and insults (PV: $LR\chi^2$ (2) $= 8.8; p = 0.01$). Ex-patients and the general population, on the other hand, are similar in the frequency with which they engage in verbal aggression. Ex-offenders also engage in more physical attacks than the other groups.
Ex-patients are less likely to give accounts (PV: $LR\chi^2$ (2) $= 10.3; p = 0.006$) or to reproach (PV: $LR\chi^2$ (2) $= 10.8; p = 0.005$) their antagonist than the other groups. This suggest that they sometimes lack the social skills necessary to engage in social control behavior. They also receive fewer accounts (PV: $LR\chi^2$ (2) $= 7.0; p = 0.03$) probably because they are less likely to reproach the antagonist.

The Order of Events

This section focuses on the position and order of actions during these incidents. Since the incidents vary in length, we coded the position of each action as a proportion of the total number of actions in the incident. For example, if there were ten actions recorded, the first act was coded 0.1, the second act 0.2, and the final act was coded as 1.0. The mean positions and standard deviations for the ten general categories of actions are presented in Table 6.3 for the four types of incidents. The actions of respondents and antagonists were not distinguished, since there was no reason to expect the position of their actions to differ.

Table 6.3. Order of actions for the three samples combined

Actions	Incidents							
	Anger		Argument		Hitting		Weapons	
	Mean	SD	Mean	SD	Mean	SD	Mean	SD
Rule violations	0.44	0.19	0.35	0.27	0.31	0.24	0.28	0.26
Orders	0.36	0.21	0.41	0.27	0.36	0.27	0.37	0.27
Reproaches	0.48	0.21	0.49	0.26	0.41	0.23	0.43	0.26
Noncompliance	0.53	0.57	0.51	0.23	0.43	0.21	0.48	0.23
Accounts	0.60	0.18	0.59	0.23	0.46	0.23	0.51	0.28
Insults	0.57	0.18	0.61	0.23	0.51	0.24	0.48	0.25
Threats	0.58	0.17	0.67	0.22	0.56	0.24	0.62	0.21
Physical attacks	–	–	–	–	0.69	0.25	0.62	0.25
Submission	0.65	0.21	0.84	0.24	0.79	0.26	0.72	0.30
Mediation	0.78	0.12	0.87	0.18	0.84	0.20	0.80	0.24

The actions are presented in the general order in which they occur, an order which is almost identical across types of incidents.[10] The order of incidents involving unexpressed anger was slightly different, with rule violations occurring slightly later. However, in general, the actions occur in the following order: (1) rule violation, (2) orders, (3) reproaches and noncompliance, (4) accounts and insults, (5) threats, (6) physical attacks, (7) submission, and (8) mediation. Thus, as expected, these incidents begin with violations of norms and orders, including orders and noncompliance, rule violations, reproaches, and accounts. Explicit attacks on identities occur later, first as insults, then as threats, and then as physical attacks. Submission and mediation does not occur until late in the series of events.

These results do not show how frequently incidents begin with violations of norms and orders. To give a rough approximation of this frequency we examined the first action in each incident to determine if it was an action related to social control (i.e., a rule violation, order, reproach, or account) or an attack (i.e.,

10 This type of analysis does not reveal sequences of events, that is, whether a given action is likely to follow other actions. Nor does it imply that each of the actions occurs in every incident.

insult, threat, or physical attack). These two categories incorporate virtually all of the codable actions in the first position. For the general population, 89% of the incidents involving unexpressed anger, 90% of the arguments, 64% of the hitting incidents, and 68% of the weapon incidents began with an action related to social control. Thus, one can conclude that about two out of three physically violent incidents and about 90% of verbal incidents begin with violations of norms and orders. The remainder begins with intentional attacks; however, the reason for these initial attacks is unclear. They could be a response to an offense that occurred prior to the incident or they could be unprovoked.

The Position of Accounts in the Social Control Sequence

It was suggested that accounts would be given in response to, rather than in anticipation of, reproaches in aggressive interactions because anticipatory accounts would prevent the conflict from developing. Thus, we expect accounts to occur immediately after reproaches, rather than immediately before or after rule violations. To determine these positions we examined the probabilities of observing the following action sequences in the data: reproach-accounts, rule violations-accounts, and accounts-rule violations. The differences between the observed and expected probabilities (based on a random model) are presented in Table 6.4. For all three samples and for all four incidents, the probability of an account following a reproach is much greater than would be expected by chance. On the other hand, the probability of accounts before and after rule violations are either negative or zero, relative to their expected probabilities. Thus the evidence suggests that accounts tend to be given in response to, rather than in anticipation of, reproaches.

The First Attack

It has been suggested that the initial attack is often an attempt to punish an offender and that subsequent attacks are retaliatory. The evidence so far shows that attempts at social control tend to occur early in the incident, but I have not shown that the person who first attacks is the agent of social control. It is possible that aggression is first used instead by targets who are resisting control or who are retaliating for the control agent's disapproval or reproach. To examine this question, it was determined whether the first attack (i.e., insult, threat, or physical attack) was carried out by the agent or target of control. Specifically, I determined who engaged in violations of rules and who gave orders during the incident and whether they were first to attack. If the social control agent tends to be the aggressor, then the actor who engaged in a rule violation should be less likely to be the aggressor and the actor who gave orders should be more likely to be the aggressor. The opposite pattern is predicted if the target of control tends to be the aggressor.

Table 6.4. Differences between observed and expected probabilities for selected action sequences

| | | Type of incident | | | | | | | | | | | |
| | | Unexpressed anger | | | Arguments | | | Hitting | | | Weapons | | |
Initial action	Following action	General population	Offenders	Patients	General population	Offenders	Patients	General population	Offenders	Patients	General population	Offenders	Patients
Reproach	Account	0.30[a]	0.24[a]	0.14[a]	0.19[a]	0.16[a]	0.17[a]	0.18[a]	0.16[a]	0.35[a]	0.16[a]	0.19[a]	0.12[a]
Account	Rule violation	−0.09[a]	0.00	−0.02	−0.03	−0.04	−0.04	0.00	−0.02	−0.03	0.00	0.00	0.00
Rule violation	Account	0.00	−0.04	0.00	−0.09[a]	0.00	−0.08[a]	0.00	−0.02	0.00	0.00	0.00	0.00

[a] $p < 0.05$.

The percentage of incidents in which the respondent engaged in the first attack is presented in Table 6.5. A comparison is made between incidents in which the respondent (columns 3 and 4) and the antagonist (columns 5 and 6) did or did not engage in rule violations and orders.

Table 6.5. Percentage of incidents in which respondent was the first to engage in an aggressive act

Type of incident	Action occurred	Respondent		Antagonist	
		Rule violation	Orders	Rule violation	Orders
Argument	No	39.1	35.8	31.8[a]	39.9
	Yes	29.0	44.9	55.2	28.9
Hitting	No	32.5[a]	25.7[a]	22.3[a]	30.2
	Yes	14.3	46.6	50.0	22.9
Weapons	No	24.8	20.8	19.0	28.2
	Yes	9.5	29.7	44.0[a]	8.9[a]

[a] $p < 0.01$.

The results suggest that the control agent is most likely to engage in the first attack. For all three types of incidents, respondents who violated a rule are less likely to be the aggressor and respondents who have given orders are more likely to be the aggressor. Similarly, for all three incidents, respondents are more likely to be the aggressor when the antagonist engaged in a rule violation, and they are less likely to be the aggressor when the antagonist gave orders during the incident. While not all of the individual comparisons are statistically significant, the same pattern is observed in every comparison.

Discussion

I have attempted to describe in some detail what occurs in aggressive encounters of different levels of severity. This sort of description is important if an interactionist theory is to be developed. Three interactionist approaches were reviewed and then used to account for the interaction patterns in these incidents. The approach that treated aggression as punishment was most useful for explaining the initial attack. About nine out of ten verbal incidents and two out of three violent incidents began with a social control process in which someone was punished for violating norms or orders. While the coercive power approach could also account for the data, it would not explicitly predict that violations are the driving force in the development of an aggressive encounter, nor does it interpret the initial attack as legitimate in some contexts and this, I believe, is crucial in explaining why it occurs. On the other hand, all three approaches help explain why retaliation occurs. Retaliation saves face, it punishes a rule violation (the original attack), and it can deter a future attack. The decision not to retaliate often reflects strategic calculations of the high cost of such behavior and thus is best explained by the coercive power approach. Thus, submission was most frequent in the more severe incidents where the costs of retaliation were undoubtedly higher.

The frequency of social control behavior in these incidents is noteworthy. For example, in verbal disputes involving the general population, reproaches and accounts were the most frequent actions. Rule violations also occurred frequently and one suspects that they would have been even more frequent if the events that occurred prior to the incident had been coded. In other words, in some cases one of the antagonists was responding to a rule violation that might have occurred earlier.

Orders also occurred with some frequency in the form of requests and commands, and these tended to result in noncompliance.[11] In addition, evidence not presented suggests that over half of the threats (56%) were contingent threats – a threat of negative consequences if the target did not comply – and thus can be considered as attempts to control the target's behavior. However, it is possible that a contingency was implied even in those threats that were coded as noncontingent.

While rule-breaking is common in everyday life, aggression is relatively infrequent. This may be because many violations go unpunished. Others are disinclined to reproach or otherwise punish a rule breaker because it creates an embarrassing scene and because they risk retaliation. Our findings are consistent with this point of view. The antagonist's rule violations were most common in incidents where the respondent was angry, but did nothing about it. This suggests that rule violations often did not result in open conflict (arguments or physical violence). When an actor did choose to respond with a reproach, open conflict was likely to occur. While reproaches focus on the target's behavior, they have indirect negative implications for the target's identity and thus lead to retaliation.

Accounts tended to be given in response to, rather than in anticipation of, reproaches. That is, accounts were not given either immediately before or after rule violations; rather, the person who violated the rule was not likely to explain his behavior unless he was challenged.[12] It was suggested that the failure to anticipate reproaches may have been one reason that conflicts occurred. If a person offers an account before being reproached, the reproach itself – which has negative identity implications for the person – is often avoided. Unfortunately, we have no data on interactions that do not culminate in anger and aggression and, therefore, we do not know how often disclaimers and other anticipatory accounts occur in nonaggressive interactions. However, everyday observation would suggest that they occur much more frequently than they do in these data.

There were strong sex differences in physical violence, but not in verbal aggression. Thus, males were more likely to use physical violence during a conflict, particularly if the antagonists were also male. This supports Feshbach's (1970)

---

11 Evidence not presented suggests that noncompliance tends to occur immediately after requests and commands.

12 Zillmann and Cantor (1976) found that accounts given for a provocation only reduced retaliation when they were given prior to the provocation; accounts given after the provocation had no effect.

argument that sex differences in aggression are due to a difference in mode of response, rather than a difference in motivational state.

While there were no sex differences observed in the frequency of verbal attack, females were actually more likely than males to engage in reproaches. Female respondents engaged in more reproaches than male respondents, and female antagonists reproached males more than male antagonists did. This could be attributed to the greater likelihood of males to engage in rule violations, but the evidence for this was inconsistent. It is more likely that reproaches are a more mild type of verbal aggression that is preferred by females. Harris (1974) found that an equal number of males and females were verbally aggressive to a man who cut in front of them in line, but that the verbal behavior of males was more severe.

The variation in the order of events makes it difficult to speak of stages in these incidents. We can identify some general patterns, however, using the data on the order of events and the Markov models. First, the social control process tends to occur early. This process involve rule violations followed by reproaches and accounts, and orders followed by noncompliance. At this point, explicit attacks begin escalating from insults to threats to physical attacks. In other words, the attacks become more serious as the incident progresses. Incidents sometimes end when one of the parties submits to the other or when third parties mediate.

In sum, this research describes the events that occur in aggressive interactions. These events are best understood as developing out of efforts to punish others for violations of norms and orders. Thus, the initial attack is justified because the target deserves punishment.[13] Targets of punishment often retaliate because they are likely to view the attack as illegitimate and thus deserving of punishment and because they wish to save face and deter future attack. These processes explain how aggressive episodes develop naturally out of other interaction processes and how they escalate.

References

Athens, L. H. *Violent criminal acts and actors: A symbolic interactionist study.* Boston, Mass.: Routledge & Kegan Paul, 1980.

Black, D. Crime as social control. *American Sociological Review.* 1983, *48,* 34–45.

Durkheim, E. *The Division of Labor* (G. Simpson, trans.). New York: The Free Press, 1947.

Felson, R. B. Aggression as impression management. *Social Psychology,* 1978, *41,* 215–213.

Felson, R. B.: An interactionist approach to aggression. In J. Tedeschi: (Ed.), *Impression management theory and social psychological research.* New York: Academic Press, 1981.

Felson, R. B. Impression management and the escalation of aggression and violence. *Social Psychology Quarterly,* 1982, *45,* 245–254.

Felson, R. B. & Steadman, H. S. Situations and processes leading to criminal violence. *Criminology,* 1983, *21,* 59–74.

---

13 One might also conceptualize the norm violator's retaliatory attack as a form of secondary deviance (Lemert, 1951) in that his aggression is a response to the social reaction he received for his initial deviant behavior.

Felson, R. B., Ribner, S. & Siegel, M. Age and the effect of third parties during criminal violence. *Sociology and Social Research,* in press.

Feshbach, S. Aggression. In P. H. Musser (Ed.), *Carmichael's Manual of Child Psychology.* New York: Wiley, 1970.

Frodi, A., Macaulay, J. & Thome, P. R. Are women always less aggressive than men? A review of the experimental literature. *Psychological Bulletin,* 1977, *84,* 634–660.

Goffman, E. The nature of deference and demeanor. *American Anthropologist,* 1956, *58,* 473–502.

Goode, W. J. Violence among intimates. In W. J. Goode (Ed.), *Explanations in social theory.* New York: Oxford University Press, 1973.

Goodman, L. A. The multivariate analysis of qualitative data. *Journal of the American Statistical Association,* 1970, *65,* 226–256.

Harris, M. B. Mediators between frustration and aggression in a field experiment. *Journal of Experimental Social Psychology,* 1974, *10,* 561–571.

Hepburn, J. R. Violent behavior in interpersonal relationships. *Sociological Quarterly,* 1973, *14,* 419–429.

Hewitt, J. P. & Stokes, R. Disclaimers. *American Sociological Review,* 1975, *40,* 1–11.

Homans, G. C. *The human group.* New York: Harcourt, 1950.

Lemert, E. M. *Social pathology: A systematic approach to the theory of sociopathic behavior.* New York: McGraw-Hill, 1951.

Luckenbill, D. F. Criminal homicide as a situated transaction. *Social Problems,* 1977, *25,* 176–186.

Luckenbill, D. F. Patterns of force in robbery. *Deviant Behavior: An Interdisciplinary Journal,* 1980, *1,* 361–378.

Miller, W. B., Geertz, H. & Cutter, H. S. G. Aggression in a boys' street corner group. *Psychiatry,* 1961, *24,* 283–298.

Rubin, J. Z. Experimental research on third-party intervention in conflict: Toward some generalizations. *Psychological Bulletin,* 1980, *87,* 379–391.

Schelling, T. C. *The strategy of conflict.* London: Oxford University Press, 1960.

Schelling, T. C. *Arms and influence.* New Haven: Yale University Press, 1966.

Scott, M. B. & Lyman, S. Accounts. *American Sociological Review,* 1968, *33,* 46–62.

Stokes, R. & Hewitt, J. P. Aligning actions. *American Sociological Review,* 1976, *41,* 838–849.

Tedeschi, J. T. Threats and promises. In P. Swingle (Ed.), *The structure of conflict.* New York: Academic Press, 1970.

Tedeschi, J. T., Gaes, J., & Rivera, A. N. Aggression and the use of coercive power. *Journal of Social Issues,* 1977, *33,* 101–125.

Tedeschi, J. T., Smith, R. B., & Brown, R. C. A reinterpretation of research on aggression. *Psychological Bulletin,* 1974, *81,* 540–563.

Zillmann, D., & Cantor, J. R. Effect of timing of information on emotional responses to provocation and retaliatory behavior. *Journal of Experimental Social Psychology,* 1976, *12,* 38–55.

Acknowledgement. This research was partially supported by Public Health Grant MH-28850 from the National Institute of Mental Health Center for the Study of Crime and Delinquency. I wish to thank Farrell Malkis and Stephen Ribner for their assistance in data analysis and Allen Liska and James Tedeschi for their comments on an earlier draft.

Chapter 7

Frustration, Aggression, and the Sense of Justice

Jorge da Gloria

Until recently the search for cognitive mechanisms mediating aggressive behavior und the external conditions under which it occurs has emphasized rational descriptive processes, neglecting the evaluative judgmental processes that may intervene in this connection. The present work will explore this line of inquiry. Most of the thinking about aggressive behavior elaborated within the realm of social psychology has tended to leave aside the problem of the functional significance of this behavior. This tendency goes back to the classic *Frustration and Aggression* (Dollard, Doob, Miller, Mowrer, & Sears, 1939), where the relationship between frustration and aggression was presented as a kind of empirical generalization based upon informal observation, and extends of the present-day theories of attribution which stress the cause-effect relationships between the events which precede the response, but fail to consider the events created by the latter. The neglect of the functional significance of aggressive behavior and the bias favoring the consideration of "objective" judgmental processes over "subjective" ones seem to be the result of the assumption on the part of the researchers that aggressive acts are pathological manifestations and, therefore, devoid of general functional significance. The historical and ideological roots of this belief, which has dominated the field of study of aggressive behavior by social psychologists, were recently highlighted in a study by Lubek (1979) who called attention to the fact that the European tradition in the study of society – represented in the United States mainly by sociologists – looked upon aggressive behavoir as serving a function in social life. It should also be underlined that biologists, starting with Charles Darwin himself, have also focused on the functions of agonistic behavior as a natural fact independent of any pathological aspect it may assume at times.

An important consequence of the particular stand taken by social psychologists about aggressive behavior was an almost exclusive limitation to the individual aspects of aggression. Despite the widely acknowledged trivial certainty that aggression is in essence interpersonal behavior, all the notions put forward by social psychologists to account for aggressive behavior fail to reflect adequately its irreducibly social nature. There have been several attempts to correct some of these shortcomings, most by biologists or by writers inspired by a biological approach. Unfortunately, many of these undertakings resulted in little more than superficial analogies between human behavior and behavior of lower animal species, overlooking the unique features of the patterns of human adaptation. When the discussion of the function of aggressive behavior turns to the consideration of the human species, the evolutionary constraint of "survival of the fittest" undergoes some inexplicable twist, and reads "survival of the strongest" in the popular versions of social Darwinism. To some authors human beings seem to escape the rule of natural law of species survival based upon some equilibrium of "hawks" and "doves," which is supposed to determine the behavior of other animal species.

A conspicuous feature of the interplay between human organisms and their environment is that to a very large extent the environment faced by the individual is the result of the actions of other individuals, past and present. The central device of the adaptation of human beings to their environment is society, and the central feature of human society is that it integrates individual learning in order to achieve collective patterns of behavior. The selective pressures of the natural environment, which in many animal species act almost directly upon the individual phenotypes, exert their action upon humans through their effect upon the outcomes of collectively patterned behavior.

If, in the case of the human species, fitness can be equated with the maximization of some quantity, this quantity cannot be the measure of some characteristic of individuals, such as strength or intelligence. Indeed, given the distinctive pattern of adaptation of the human species, based upon the collective transformation of the natural environment, fitness has to be understood in this case as referring not to individual characteristics, but to group properties, or, more precisely, to patterns of interindividual behavior arising consistently out of the pairing of given individuals. Of course, the difference between the patterns of adaptation of the human species and those of other animal species is a matter of degree, more than a matter of nature. Individual learning of collective patterns of action is documented in several species, from birds to primates. Social diffusion of individual technological discoveries has been noticed in birds and primates, and variations in the social organization of groups of the same animal species living in distinct environments have been repeatedly studied. However, the specific combination of these different mechanisms, as well as their relative importance in the general economy of the survival of the species, seem typically human.

The kind of adaptative mechanisms that seem to be at work in human species rules out the possibility of achieving the coordination of the actions of individuals through an extended system of fixed patterns of action which would be brought into play by specific releasers. The varied environments to which human groups have to adapt, and the unique patterns of change that are followed by these environmental adaptations as the result of human activity, call for a control of interindividual behavior which has to be both reliable and flexible. Besides providing direct means of survival for the individual organisms, human societies have to develop and maintain means of achieving control of interindividual behavior which, being themselves preconditions for accomplishing the main function of the societies, acquire a survival value for the individual members of the society.

However, if we were to draw conclusions from the survival value of the control of interindividual behavior and apply them to the determination of this latter process by principles of natural evolution, this would be a hasty jump. As any other process of control, the control of interpersonal behavior requires both a specification of the set of the input-output relationships which are to be maintained and the available physical devices performing this function. The genetic endowments of the individuals provide means of control of interpersonal behavior, but define the sets of input-output relationships to be maintained in terms of the individual organisms, not in terms of interindividual activity. Nature tells each of us that some external condition is injurious or lethal, but leaves us to find how to use our abilities in order to bring other members of the species to contribute to spare us this condition. Therefore the core of the system controlling interpersonal behavior is some means of achieving appropriate definition of the set of acceptable input-output relationships and of the individual acts appropriate for maintaining external events within this set. We suggest that justice, understood in a broad sense to be qualified later, is the major component of this control mechanism.

The effective performance of this control system rests upon efficient transmission of information between individuals and adequate storage and treatment of the information transmitted. The first function is accomplished either through the triggering by specific releasers of fixed patterns of action consistently awakening particular responses or by decoding and encoding the behavior of other agents. For the reasons already advanced, we consider that the first of these processes plays a minor role in the adult individual, where the releasing of fixed action patterns comes under the control of cognitive processes for the most part. The second mechanism intervening in the control of interpersonal behavior, encoding and decoding the behavior of social actors, rests upon the same general cognitive abilities which contribute to delimit the set of behaviors acceptable from different sorts of agents, and to store this knowledge in the form of normative prescriptions. The interdependence of these two functions, building up normative prescriptions concerning the behavior of members of a society, and encoding and decoding their behavior, probably accounts for the usually found contamination of causal judgments by the ethical beliefs of the judge. There exist in principle two different methods for approaching the study of normative systems. One consists in studying, by pure observation or by experiment, naturally evolved norms; the second goes through the elaboration of theoretical principles leading to hypotheses relative to the principles which may give rise to different norms which will be experimentally created in order to be studied. For the most part this later approach will be followed here.

We consider that norms arise in a given situation as products of inferential reasoning. Features of the physical and social environment and abilities and dispositions which the individual attributes to another agent lead the individual to consider some reactions of the other agent as being logically necessary; for instance, if the individual knows that another will perform a given act in order to obtain a given amount of a reward, he will consider, as a necessary consequence, the performance of the same reaction when it commands a larger amount of the same reward. This kind of calculation affords the individual some knowledge of the forthcoming acts of potential partners. A very important feature of these calculations is that they take into account not only the idea which the individual makes of himself and of his own abilities, but also the way in which he thinks that his potential partners see him and his abilities. Performing this kind of calculation enables the individual to draw conclusions concerning the way other individuals should act towards the individual as he sees himself.

Facing an eventual partner, the individual will appraise whether the results of the behavior to be expected on his part are more favorable to the individual than would be the results of acting alone or with another partner, and, on the basis of his appraisal, the individual will accept or refuse to interact with a potential partner. In the first case, the behavior of the partner predicted by the individual will appear to the latter as being not only logically necessary or "reasonable" but also as "just." The latter is a direct consequence of the way in which the individual achieves the prediction of the behavior of potential partners, taking into account their preferences and abilities, as well as external conditions, including himself, in order to ascertain the probable course of their action. In this way it is ensured that the behavior predicted by the individual is the one which maximizes the outcomes for both partners under the available conditions.

The same principles which account for the rise of a system of reciprocal obligations between social agents constitute the basis of the mechanisms which encode and decode social behavior.

The individual who notices an external event that contradicts his current assumptions concerning his environment will check the features of the event he noticed against alternative formulations of these assumptions. Starting with the more probable explanations, he will move on to less probable ones, until he finds a sufficient reason for the event he noticed. By sufficient reason we mean a set of environmental conditions, including the preferences and abilities of other agents, through which the individual presumes to control some of the features of the environment, such that the event to be explained could logically result from the joint action of these factors. The outcome of this process will be the rejection of some of the assumptions on which the prediction originally made by the individual was grounded, resulting in the redirection of his activity along new lines consistent with his revised perception of the situation.

Two features of this process should be kept in mind in order to study it in concrete cases: first, the individual tests some specific hypothesis, referring to a well-identified factor of factors, against some unspecified null hypothesis, referring to other factors; second, the subject is not seen as attempting to reject that unspecified null hypothesis, but as trying to find confirmatory evidence for the acceptation of his specific hypothesis. For this reason, attributions made by individuals seem to lack the kind of binary symmetry which we too often seem to assume. For instance, subjects will try to find out whether a given act was intentionally performed and, if they cannot find confirmatory evidence for this hypothesis, will simply try to find confirmatory evidence for some other specified hypothesis, but not perform under the acceptation of the unspecified attribution of "unintentional act."

Application of the principles which we have been outlining to the study of aggression is less straight forward because of the confusion and the oversimplifications reigning in the delimitation of this process. Every observable act constitutes at the same time a *reaction,* determined by events preceeding the performance of the act, and an *action,* a set of organized reactions tending to achieve some result, determined by the current perceptions of the situation in which the subject is acting. The behaviors most often studied in social psychological experiments – *revenge, demonstrations of force, defense* – constitute instances of *actions* and not of *reactions.* The latter have been little studied, except for their neurovegetative parameters, studies of neuromuscular activity being conspicuously scarce.

The separation of these two aspects of behavior leads to the formulation of distinct hypotheses at each level. For instance, while *reactions* to frustration will be expected to reflect directly the intensity and severity of the frustrating event, posterior *actions* will translate the qualitative nature of the frustrating event in the attempts made by the individual to find the course of action appropriate to overcome the frustration. We have no reason to expect that, in general, a more severe frustration will produce, as an exigence of appropriate action, a more aggressive behavior on the part of the victim.

The qualitative characterization of the situations, which will determine the course taken by the actions of the individuals, will result from the cognitive operations performed by the individual based upon the information available to him. Two related sets of operations have to be considered in this connection: one which identifies the features of the event and determines the *reaction* of the individual and another which takes into account the nature of the event, the external circumstances, and the abilities of the individual in order to control the *action* appropriate to the event noticed.

The situations which have been studied as contexts of aggressive interactions do not call for an exhaustive discussion of attribution processes, centering around a small set of supposedly critical features of these situations. We will therefore concentrate on an analysis of some selected topics.

A close relationship exists between the prediction of external events and goal-directed activity. At a purely cognitive level, a goal is nothing more than the prediction made by an individual concerning the outcomes of his own behavior. Failure of a prediction of this kind constitutes, therefore, a frustration in the sense of the "frustration-aggression hypothesis." A major source of information concerning a frustrating event is, therefore, the prediction originally made by the individual. The way in which the current events differ from his prediction tells the individual which factors among those originally considered failed to behave in the predicted way. This first step will lead him to conclude that the part of his original prediction disconfirmed by the observed event concerns either his own behavior or some condition external to the individual. In the latter case, further analysis will tell the individual whether some feature of the physical environment or some act of another individual differs from those originally considered.

A first qualitative classification of the frustrating event will result from this analysis: the event noticed constituted a *failure* of the individual to behave in some way originally taken to be appropriate; it constituted an action of some other individual or resulted from some feature of the physical environment.

Most research concerning aggressive behavior has studied the effects of frustrating events due to actions of others, that is, either events arising in the context of an action undertaken by the individual – *obstructions* – or in the context of action undertaken by others – *attacks*.

We shall consider separately failures on the one hand and obstructions and attacks on the other. Failure directly denies the possession of some ability which the individual thought he possessed, while obstructions and attacks specifically deny the ability to control the behavior of another to the extent previously thought by the individual. Indeed, as we have seen, the individual being himself one of the external determinants which shape the actions of others, obstructions and attacks mean that the victim lacks the abilities to support the preferences which are contradicted by the actions of the attacker.

The basic proposition, the validity of which we will try to establish, asserts that social actors react to events which imply a reduction of their sense of worth by performing acts which, under the conditions at hand, will restore their self-evaluation. This proposition applies to aggressive behavior, as far as this appears to the individual as a means of asserting his worth. We do not affirm that individuals will react aggressively in all instances in which their sense of worth is diminished, and we do not want to imply that all aggressive behavior procedes from attempts of the individuals to restore their self-worth.

The simple principle we have just advanced seems to account for a large body of experimental results. We will review some of these results, first those which illustrate the reactions of individuals to failure, attack, and obstruction and then turn to those which illustrate the different actions of subjects in coping with failure on the one hand and with attacks and obstructions on the other. Finally, we will discuss the relationships between actions aimed at coping with different aversive events and reactions to such events.

Reactions to Failure, Obstruction, and Attack

The theoretical analysis which we have presented leads us to formulate distinct hypotheses concerning the *reaction* to frustrating events and the *actions* of the individuals who are the victims of such events. The frustration-aggression hypothesis makes the same prediction concerning the behavior of a victim of frustration, whether the frustration appears to the individual as being the product of his own behavior or as resulting from some other cause. The basis of this prediction is the belief in the determination of the reaction of the individual by the objective character of the external stimulation.

We consider this prediction to be justified as it applies to the reaction of individuals to frustrating events, but we reject it as a prediction of the acts of the victims of these events.

Except for a few studies of effects of frustration upon muscle activity (Haner & Brown, 1955; Roehl, 1960), most of our data on the reaction of human subjects to attack and failure concern neurovegetative arousal. The principles we have outlined lead us to predict that failure, as well as attack, should result in autonomic arousal. In the case of failure, we would predict an increase in autonomic arousal proportional to the severity of the frustration undergone by the subject. In the case of frustrations attributable to the action of another agent, we would predict a level of arousal determined by the sum of the severity of the frustration and the extent to which the obstruction to the activity of the subject threatens the self-evaluation of the latter. An identical prediction would be made in the case of attack: we would expect the reaction of the individual to reflect both the severity of the consequences of the attack and the specific motives attributed to the attacker. A very large number of experiments, having used quite distinct operationalizations, shows that attack or insult directed against an individual results in increased autonomic arousal. This effect was found with systolic blood pressure (Hokanson & Edelman, 1966; Kahn, 1966), diastolic blood pressure (Frodi, 1978; Gambaro & Rabin, 1969), vasomotor reaction (Stone & Hokanson, 1969), and skin conductance (Taylor, 1967). Failure at some task which threatens the self-evaluation of the individual also entails a reliable effect upon the state of autonomic arousal of the victim, analogous to the effect produced by attack. For instance, Gentry (1970b) found a difference in systolic and diastolic blood pressure after failure at an intelligence test or insult by the experimenter; Spiegel and Zelin (1973) found an analogous difference after failure at a test of arithmetic ability. Several experiments performed by Hokanson and his associates show reliable increases in systolic blood pressure after an intervention by the experimenter to give the subject the impression that he is performing a counting task poorly (Hokanson & Shetler, 1961; Hokanson & Burgess, 1962; Hokanson, Burgess, & Cohen, 1963).

Available experimental evidence is not conclusive about what concerns comparison of the levels of arousal induced by failure, obstruction, and attack, though some data would point to the fact that insult or attack might add their effects to the effects of failure (Rule & Hewitt, 1971). One experiment, showing that failure and insult result in a smaller increase of diastolic blood pressure than success and insult (Gentry, 1970a), tends to demonstrate, in spite of difficulties of interpretation of the measurements performed, that justifiability of the insults against the subject tends to reduce their effects upon arousal. This result is consistent with the hypothesis of an arousing effect of interpersonal treatment tending to devalue the subject. Should this hypothesis be true, one would expect that insulting remarks coming from an irresponsible individual would be less arousing than the same remarks coming from a normal partner. Such effect was found by Zillmann and Cantor (1976) who studied the blood pressure of individuals insulted by normal and emotionally upset partners. Thus, our brief survey of the experimental literature leads to the conclusion that measures of autonomic arousal directly reflect the extent of threat to self-evaluation which a given experimental treatment constitutes.

Results from another line of research, followed by J. Hokanson and his associates, illustrate the means whereby the autonomic balance disturbed by failure and attack returns to its chronic state. Several experiments show that performance by the subject of acts consistent with both his role and the causal structure of the situation, tending to bring to an end the aversive experience of failure or attack, restore the autonomic balance of subjects (Hokanson & Edelman, 1966; Hokanson, Willers, & Koropsak, 1968; Stone & Hokanson, 1969).

Coping with Failure

Stressing, as we have done, the role of the judgment made by the individual of the situation which faces him, as well as his judgments of the different courses of action which he considers, we are led to predict different *actions* from the victim of frustrations which appear to the individual as being the result of his own action and to frustrations which seem to result from the action of other actors. Moreover, in this latter case, we predict reactions which will depend on specific representations of the motives of the other actor elaborated by the individual. In the case of failure, i. e., frustrations attributable to the individual's lack of ability, we will not predict, in general, an aggressive reaction but, instead, an attempt to demonstrate competence or to escape from the situation.

Some of the experiments setting out to demonstrate the frustration-aggression hypothesis and obtaining negative results do, in fact, confirm our assertion concerning the reactions to failure. For instance, Buss (1963) found a significant difference between the amount of shock delivered in his standard learning situation by subjects who were led to believe that they had failed their training of the confederate and by those who thought that they were performing that task successfully. In a later replication of this experiment, the author (Buss, 1966) created two different conditions by telling half of his subjects that the stronger the shocks delivered, the faster the learning by the confederate; no mention of the effect of the shocks was made to the other half of the subjects. In this new experiment the difference previously found occurred only in the groups of subjects which had been told that stronger shocks facilitated learning, showing that the effect of failure in the training task was an increase in the attempts to train the confederate and not an increase in aggressive behavior.

The well-known experiment performed by Geen (1968) illustrates the same fact. In this experiment, the subjects try to train a confederate of the experimenter in a learning situation similar to the one used by Buss. In the first phase of the experiment, the subjects are instructed to punish the errors made by the confederate by administering to the latter an electric shock, and to reinforce the correct responses by flashing a light. Half of the subjects were verbally reinforced by the experimenter whenever they delivered a shock. During the interval between the first phase and the next identical phase, the subjects were exposed to one of the following experimental treatments: failure at an intelligence test; obstruction on the part of the confederate, preventing the subject from successfully terminating an intelligence test; insult by the confederate; no intervention (this control group was allowed to complete the test without interference or insult from the confederate). Obviously, the experimental procedure just described does not allow a clear-cut separation of aggression from attempts to achieve success in the learning task. However, from the published results, some tentative conclusions can be drawn. For instance, increase of the intensity of shocks intended to promote learning will not occur before the confederate has made some errors, while we would expect that increase of shocks expressing aggression would manifest itself as soon as the subject is given the opportunity to deliver shocks. In fact, in the first block of trials, subjects who failed in the intelligence test did not differ from controls, but subjects who were insulted or the victims of obstructions did differ from controls in this block of trials. If the increase of shocks constitutes an attempt to promote learning, and not an expression of aggression, we would expect that the subjects having been reinforced after sending shocks would increase their administration of shocks as soon as they discover that the confederate does not learn, while we would expect those who were not reinforced to increase or decrease the intensity of their shocks in a quite random fashion. As predicted, subjects who failed the intelligence test and were led to believe that higher shocks facilitated learning differ from controls from after the second block of trials through the last block of trials, while those who do not hold this belief differ from controls in the second and fourth blocks of trials, but not in the first and third blocks. The results obtained by Geen (1968) point to the conclusion that failure of the subject in performance of one task has led him to try to perform successfully another task, while deliberate obstruction, as well as insult on the part of another, led him to behave offensively towards this other agent.

Other experiments which utilized different operationalizations (Taylor & Pisano, 1971) lead to exactly the same conclusion. No experimental result known to the author demonstrates the occurrence of aggressive behavior following failure at a nonaggressive task. Despite its reliable effect upon autonomic arousal, failure induced the subjects to engage in the usually appropriate behavior necessary to succeed in the task, though the experiments which were performed have not studied any condition under which an inappropriate activity (for instance, aggression) could be interpreted by the subjects as asserting their own personal worth.

Coping with Obstruction and Attack

While our review of the literature points to the conclusion that failure does not, in general, foster aggressive behavior, a large amount of data demonstrates that both obstruction and attack launched by another will have this effect. For instance, the experiment by Geen (1968) shows that subjects whose performance in the intelligence test suffered the interference of the confederate delivered shocks almost as intense as those delivered by insulted subjects, both groups delivering shocks more intense than those delivered by the control group. Gentry (1970a) also showed that insulted subjects will deliver more shocks and stronger shocks than controls.

Despite the wide agreement of the data on this point, there are large variations in the amount of aggression which will be exhibited by victims of attacks or obstructions on the part of another individual, depending on the circumstances. A large number of features of obstructions and attacks have been presumed to account for these variations: these acts may appear to the victim as being reasonable or not (Pastore, 1950), as arbitrary or justified (Pastore, 1952); they may or may not be expected by the victim (Berkowitz, 1960a; 1960b); they may seem to be intended or not Greenwell & Dengerink, 1973); they may or may not be felt as constituting a violation of rights (da Gloria, 1975).

Each of these conditions, separately considered, was shown in more than one experiment to have an effect upon the level of aggression exhibited by the victims of attacks or obstructions. A few experimenters attempted the simultaneous manipulation of more than one of the factors indicated in order to detect possible interactions. A well-known experiment of this kind was performed by Kregarman and Worchel (1961) and is considered by these authors to show that arbitrary frustrations induce more aggressive responses on the part of the victims than frustrations which appear to be justified, because arbitrary frustrations are usually unexpected to the victims. However this conclusion does not seem to be acceptable given the flaws in the conception of the experimental treatments and the fact that the differences among these treatments found by this authors are not, contrary to what they state, statistically significant. Commenting upon the difficulties they found in their attempts to design independent manipulations of arbitrariness and expectedness, Kregarman and Worchel note that a justified frustration is usually not unexpected, though an arbitrary frustration can be expected as well as unexpected. The difficulty met by Kregarman and Worchel seems to reflect the shortcomings of the discussions of this question which go back to the original paper of Pastore (1950) preceeding the better-known experimental paper published by the same author in 1952. In the first of these papers, Pastore considers several aspects of potentially frustrating situations: reasonableness and justification. In his latter discussion of the problem, justification, being paired with arbitrariness as its contrary, loses its original connotation of justice, and comes to mean merely that the action which is not arbitrary results from the effect of some constraint external to the agent. The notion of expectedness comes in in this context, denoting a project of action in the writing of Pastore, and the subjective probability of an event in the writings of Berkowitz (1960a, 1960b). In most cases these authors relied upon intuition and common sense to provide precise boundaries for these terms. However, it seems that this resulted in a confusion among attributions, subjective reconstructions of the motives of others, and objective descriptions of the features of situations. As applied to the notion of expectancy, this approach leads one to consider as identical what is expected because it is required and what is expected because it is probable, thus confounding probabilities and norms. It seems possible to suggest a more systematic treatment of this question. *Reasonableness* and justification are in one sense two different terms to say the same thing: a given event results from the activity of another agent who is controlled by some predictable feature of his environment. In this sense a justified or a reasonable frustration is simply a frustration which is not attributed to the author of the act that engendered the frustration. Such an act cannot communicate to the victim any evaluation of his abilities on the part of the author and should not be expected therefore to result in aggression towards the latter. A closely similar situation arises with acts sometimes called "unintentional." To study this kind of act, the aversive event is presented to the victim as being the result of some factor, external to both the victim and the author, which the latter does not control. This kind of act should not, for identical reasons, lead to aggressive actions. That acts which are devoid of sense to the victim because they do not originate in another individual appear as more or less probable should not affect the former's behavior. In another sense, a justified frustration is one which appears to the individual as resulting from an act which is consistent with the individual's sense of what is "just" in the situation. Such an act is by nature reasonable, since justice reflects the order of things as it is perceived by the individual. Though such an act may be considered as being more or less probable, it is always taken to be an obligation and as such cannot be unexpected without ceasing to be justified.

Many experimental data show that acts which do not appear to the victim as originating in another individual do not lead to aggression. For instance, Kregarman and Worchel (1961) have shown that obstructions resulting from some unavoidable deficiency of the author do not entail hostile reactions. Greenwell and Dengerink (1973) have also shown that the level of shock set by the attacker in a shock box experimental situation – and not the level of the shocks really administered – was the critical factor of the retaliation by the victim. Nickel (1974) draws the same conclusion from the experiment he performed.

These results are often taken to mean that an aggressive reaction of the victim of an attack or a frustration necessitates that the act generating the frustration or the attack is perceived as being intentional. Quite probably such a view is not correct, though it has been so generally held that nobody seems to have attempted to study aggressive behavior aimed at authors of unintentional aversive actions. However, such behavior seems to be frequent in natural conditions, arising in the context of work and traffic accidents, when clearly aggressive actions are directed against the careless driver or the greedy plant owner.

Thus the role of intentionality is not a simple one: unintentional acts per se cannot convey information about the motives of the agent and therefore are not able to express the latter's contempt for the victim; however, if some feature of the situation in which an unintentional act occurs makes its performance able to communicate the motives of the agent, the act expresses contempt and will lead to aggressive behavior whenever such behavior appears to the individual as appropriate in order to assert his worth.

Of all features of a situation of interaction, the one which most directly provides the victim of a given unintentional act with an insight on the value that the performer of the act puts upon the satisfaction of the victim's preferences is the latter's previous commitment to the fact that this given act will not be performed. Therefore, whenever an agent implicitly or explicitly accepts not to perform a given act resulting in some event in exchange for the performance of some act by another agent, the simple occurrence of this event, whether resulting from an in-

tentional act or from an unintentional one, will express the contemptuous neglect of the preferences of the other agent. Indeed, the original agreement upon
which the victim set up his course of action was based upon the presumed ability
of his partner to prevent the occurrence of the undesirable event and assumed
the effective exercise of this ability. Besides providing an explanation for aggressive behavior in response to unintentional performance of aversive acts, the former argument gives some clues concerning the assessment of the degree of contempt to be perceived in the performance of a given aversive interpersonal act:
the less this act will seem to profit the agent, or the more difficult it seems to perform, the more the contempt on the part of the performer will be resented, for a
given level of experienced aversiveness.

Several experiments have attained results consistent with the principles outlined. One illustrates the case of violation of explicit contracts; some others
study the infringement of norms embedded in implicit contracts. A well-known
experiment performed by Worchel (1974) constituted originally an attempt to
study the effects of randomness, expectedness, and justification of a frustration
upon the level of hostility subsequently manifested by the victims towards the
frustrator. This author interprets the results of his experiment in terms of "reactance" along the lines suggested by Brehm (1966). However, several features of
the data do not fit in this interpretation: in subjects who were barred from the
choice of the reward, after having been told they would be offered such a choice,
hostility increased as the desirability of the reward assigned to them by the
experimenter decreased. In his discussion of the frustration-aggression hypothesis, Brehm (1966) explicitly rules out the interpretation of this type of event in
terms of "reactance." He states that the amount of "reactance" is a function of
the initial desirability of the suppressed alternative, but the behavior of the
"reactant" subject is determined by the chances afforded by the available
courses of action of overcoming the restrictions imposed on his choices. It is possible to suggest a different interpretation of this data: the three experimental
conditions studied by Worchel illustrate different types of implicit contracts between the experimenter and the subjects which are, or are not, honored later on.
Subjects in this experiment reacted aggressively to the violation of the contract
on the part of the experimenter each time that this violation could not be ac-
counted for by external conditions acting upon the frustrating agent. When the
experimenter tells the subject that the experimenter will choose the reward to be
given to the subject and later on gives the latter one of the rewards, telling him
that the subject will get that precise reward because the experimenter has decided to give the same number of each type of reward, no contract is broken
whatever reward the subject receives. In this condition, subjects express little
hostility, and they express the same level of hostility whatever their preference
for the reward they received. When the experimenter tells the subject that he will
be free to choose the reward that he will be given and later on breaks this contract, subjects manifest great hostility, even in the case where they are given their
preferred reward. The degree to which the preferences of the subject are neglected by the experimenter, without reason, constitute for the former an indication of
the ill will of the experimenter toward him and lead, therefore, to an increase in

hostility proportionate to the degree of nonsatisfaction of the preferences of the subject. Finally, the subjects in the group to which the experimenter has promised that they will be given the reward previously designated by chance will see this promise broken with a justification grounded on a principle analogous to the previously accepted random assignment. In this case, subjects who receive the most desirable reward will see the contract with the experimenter fully honored; those given the least desirable reward can attribute their dissatisfaction to a shortage of the reward they were to receive, as well as to the ill will of the experimenter who could perhaps have given them their second choice. Those given their second choice will only notice that the experimenter did his best to satisfy their preferences, given the rewards available. In accordance with this explanation, subjects given their second choice manifest slightly less hostility than those who were fully satisfied; those given their last choice express a degree of hostility intermediate between the subjects whose choice was barred and those given their least preferred choice after the experimenter led them to recognize his right to allocate the rewards.

The effects of the violation of implicit contracts, as well as the effects of perceived outcomes of an aggression as indicators of its seriousness, were studied in several experiments performed by this author and associates.

In one experiment, we studied the establishment of norms and the responses to their infringement in pairs of subjects. Both subjects were instructed to perform a motor task and their success in this task was rewarded with money. The ostensible goal of the experiment being the study of motor coordination under conditions of unpredictable stress, each of the subjects was instructed to deliver an electric shock to the other during the latter's execution of the motor task. Successful obstruction of the performance of the motor response was also rewarded. The subjects were told that a successful obstruction depended on proper timing of the shocks, as well as on their intensities. Three different intensities were made available, and the subjects were led to believe that administration of the weaker shocks, given appropriate timing, resulted in a successful obstruction half the time, and both the strongest shock and the intermediate one always resulted in a successful obstruction if delivered at the proper time. The shocks were, in fact, delivered by the experimenter in such a way that whenever a subject received the stronger shock, he got a shock of the same intensity of the intermediate shock administered to the other subject.

The results of this experiment show that the critical factor determining the behavior of the subjects was the idea that the intensity of the shock they received was the one required for the successful performance of the motor task by the other subject. When they received the supposedly intermediate shock, the subjects would deliver a shock of intermediate intensity; when receiving the supposedly stronger shock, they responded with the stronger shock, despite the fact that the intermediate shock and the stronger one were identical. In one case, the subjects who received the intermediate shock responded with the stronger one: this happened when the profit of the attacker did not appear to the victim as being a sufficient reason for the pain inflicted. Whenever the subject administering the intermediate shock was known to attain a reward whose value, compared with the value of the reward lost by the victim, was insignificant, the latter responded with the stronger shock.

Several experiments in different natural populations and using different operationalizations of the notions outlined replicated these results.

Besides its role in the control of aggressive behavior, the principle of sufficient reason was found to be an important determiner of the judgments of crimes (da Gloria & Duda, 1980). In several experiments we gave written accounts of two kinds of homicides to populations of students. In one case a robber kills the owner of a given sum of money in order to rob this sum; in the other case, the owner of the sum kills the robber. The episodes give three different values to the sum of money: 5000, 100000, and 250000 French francs. Subjects are required to rate the seriousness of the homicide. Their ratings show that the homicide committed by the robber is considered to be more serious than the one committed. by the owner of the sum. In both of these cases, the higher the sum at stake, the less serious the homicide was considered to be.

From our survey of the literature, a quite manageable account of one aspect of the dynamics of aggression seems to emerge. Individuals experience aversive events, some of which they attribute to the activity of other social agents. Some of this category are construed by the victims to mean that other social agents see them as being less valuable than they think themselves to be, and this constitutes a supplementary aversive experience. They react to the aversive experiences by a degree of autonomic arousal which corresponds to the degree of aversiveness of their experiences, and they behave in ways usually appropriate to cope with the aversive event as they construed it: demonstrating competence whenever they fail in a task, asserting control upon the behavior of the other whenever the latter has questioned this ability. Performance of the appropriate behavior results in the confirmation of the individual's abilities and restores the autonomic balance.

References

Berkowitz, L. Some factors affecting the reduction of overt hostility. *Journal of Abnormal and Social Psychology,* 1960, *60,* 14–21. (a)

Berkowitz, L. Repeated frustrations and expectations in hostility arousal. *Journal of Abnormal and Social Psychology,* 1960, *60,* 422–429. (b)

Brehm, J. W. *A theory of psychological reactance.* New York: Academic Press, 1966.

Buss, A. H. Physical aggression in relation to different frustration. *Journal of Abnormal and Social Psychology,* 1963, *67,* 1–7.

Buss, A. H. Instrumentality of aggression, feedback, and frustration as determinants of physical aggression. *Journal of Personality and Social Psychology,* 1966, *3,* 153–162.

da Gloria, J. *Les déterminations cognitives des conduites agressives analyse expérimentale de la régulation normative de l'interaction.* Unpublished doctoral dissertation, Université Paris 7, 1975.

da Gloria, J., & Duda, D. Conditions de la légitimation de l'auto-défense. Etudes préliminaires. *Recherches de Psychologie Sociale,* 1980, *2,* 3–26.

Dollard, J., Doob, L., Miller, N., Mowrer, O. M., & Sears, R. R. *Frustration and aggression.* New Haven: Yale University Press, 1939.

Frodi, A. Experiential and physiological responses associated with anger and aggression in women and men. *Journal of Research in Personality,* 1978, *12,* 335–349.

Gambaro, S., & Rabin, A. K. Diastolic blood pressure responses following direct and displaced aggression after anger arousal in high- and low-guilt subjects. *Journal of Personality and Social Psychology,* 1969, *12,* 87–94.

Geen, R. G. Effects of frustration, attack and prior training in aggressiveness upon aggressive behavior. *Journal of Social Psychology.* 1968, *9,* 316–321.

Gentry, W. D. Effects of frustration, attack and prior aggressive training on overt aggression and vascular processes. *Journal of Personality and Social Psychology,* 1970, *16,* 718–725. (a)

Gentry, W. D. Sex differences in the effects of frustration and attack on emotion and vascular processes. *Psychological Reports,* 1970, *27,* 383–390. (b)

Greenwell, J., & Dengerink, H. A. The role of perceived versus actual attack in human physical aggression. *Journal of Personality and Social Psychology,* 1973, *26,* 66–71.

Haner, C. F., & Brown, P. A. Clarification of the instigation to action concept in the frustration-aggression hypothesis. *Journal of Abnormal and Social Psychology,* 1955, *51,* 204–206.

Hokanson, J. E., & Burgess, M. The effects of status, type of frustration and aggression on vascular processes. *Journal of Abnormal and Social Psychology,* 1962, *65,* 232–237.

Hokanson, J. E., Burgess, M., & Cohen, M. E. Effects of displaced aggression on systolic blood pressure. *Journal of Abnormal and Social Psychology,* 1963, *67,* 214–218.

Hokanson, J. E., & Edelman, R. Effects of three social responses on vascular processes. *Journal of Personality and Social Psychology,* 1966, *3,* 442–447.

Hokanson, J. E., & Shetler, S. The effect of overt aggression on physiological arousal. *Journal of Abnormal and Social Psychology,* 1961, *63,* 446–448.

Hokanson, J. E., Willers, K. R., & Koropsak, E. The modification of autonomic responses during aggressive interchange. *Journal of Personality,* 1968, *36,* 386–404.

Kahn, M. The physiology of catharsis. *Journal of Personality and Social Psychology,* 1966, *3,* 278–286.

Kregarman, J. J., & Worchel, P. Arbitrariness of frustration and aggression. *Journal of Abnormal Social Psychology,* 1961, *63,* 183–187.

Lubek, I. A brief social psychological analysis of research on aggression in social psychology. In A. Buss (Ed.), *Psychology in social context.* New York: Irvington, 1979.

Nickel, T. W. The attribution of intention as a critical factor in the relation between frustration and aggression. *Journal of Personality,* 1974, *42,* 482–492.

Pastore, N. A. A neglected factor in the frustration-aggression hypothesis: a comment. *Journal of Psychology,* 1950, *29,* 271–279.

Pastore, N. A. The role of arbitrariness in the frustration-aggression hypothesis. *Journal of Abnormal and Social Psychology,* 1952, *47,* 728–732.

Roehl, C. A. The effects of frustration on the amplitude of a simple motor response. *Dissertation Abstracts,* 1960, *20,* 4188.

Rule, B. G., & Hewitt, L. S. Effects of thwarting on cardiac response and physical aggression. *Journal of Personality and Social Psychology,* 1971, *19,* 181–187.

Spiegel, S. B., & Zelin, M. Fantasy aggression and the catharsis phenomenon. *The Journal of Social Psychology,* 1973, *91,* 97–107.

Stone, L., & Hokanson, J. E. Arousal reduction via self-punitive behavior. *Journal of Personality and Social Psychology,* 1969, *12,* 72–79.

Taylor, S. P. Aggressive behavior and physiological arousal as a function of provocation and the tendency to inhibit aggression. *Journal of Personality,* 1967, *35,* 297–310.

Taylor, S. P., & Pisano, R. Physical aggression as a function of frustration and physical attack. *Journal of Social Psychology,* 1971, *84,* 261–267.

Worchel, S. The effect of three types of arbitrary thwarting on the instigation to aggression. *Journal of Personality,* 1974, *42,* 300–318.

Zillmann, D., & Cantor, J. R. Effect of timing of information about mitigating circumstances on emotional responses to provocation and retaliatory behavior. *Journal of Experimental Social Psychology,* 1976, *12,* 38–55.

Chapter 8

The Relations Among Attribution, Moral Evaluation, Anger, and Aggression in Children and Adults

Brendan G. Rule and Tamara J. Ferguson

Retaliation by a victim, or punishment by law for an act of harm, requires judgment about the motives underlying the act, its avoidability, and the amount of harm done within a normative context (Feshbach, 1971). However, despite the recognition by several authors (Feshbach, 1971; Pepitone, 1976, 1981; Tedeschi, Smith, & Brown, 1974) that such factors contribute to an understanding of reactions to harm, relatively few analyses of aggression have explicitly incorporated these cognitive and normative considerations.

The purpose of this chapter is to specify how normative and attributional factors combine to affect reactions to, and evaluations of, a harmful act. More specifically, we wish to specify the conditions under which people respond to harmful incidents *as if* they were aggressive ones. Our focus, therefore, is not on whether people consciously label an act as aggressive, but whether the act arouses emotions and judgments that *might* lead to the labelling of an act as aggressive. We shall deal with several issues. First, we shall identify the bases for causal assignment and discuss the relation between such assignment and norms. Second, we shall review the data showing that people do react to harm violations as we propose. Third, we shall review the conditions under which biases in cognitive processing are evident. Finally, we shall draw some conclusions.

The Role of Norms and Causality in Blame, Anger, and Aggression

We, among others (such as da Gloria, 1977; Mummendey, Bornewasser, Löschper & Linneweber, 1982; Pepitone, 1976, 1981; Tedeschi et al., 1974) recognize the importance of normative beliefs in regulating people's reactions to harmful incidents. Such normative beliefs or "oughts" refer to the imperativeness aspect of the obligation to act, think, or feel in certain ways (Hollingsworth, 1949), and can be either prescriptive or proscriptive in nature. One class of normative beliefs has been discussed in detail by Pepitone (1976, 1981). These are beliefs regarding property rights, human welfare, and the social order, as well as fairness and justice. In our view, violation of these normative beliefs contributes to whether an event is seen as harmful in the first place. Pepitone's analysis stops, however, at the point where observers or victims perceive that another person is actually causally responsible for such norm violations and are thereby instigated to aggress. We think, however, that a causal analysis of the situation continues. Thus, another kind of normative belief can be identified. These beliefs embody attributionally relevant considerations of the way in which a person should or should not be personally or causally responsible for behavior, including harmful behavior.

Our analysis thus begins in essence where Pepitone's analysis ends. We begin with the assumption that a perceiver views an event as a harmful one. The question of interest to us is how the actor's causal responsibility for harm adds to whatever negative reactions may be aroused by the simple fact that harm has occurred or could have occurred. It is at this point where the norms of proper conduct we have distinguished become important considerations. The norms of proper conduct we describe were derived from Heider's (1958) analysis of personal responsibility (see also Fishbein and Ajzen, 1973) and are defined in Table 8.1. On the left panel, we have listed Heider's five levels of personal responsibility, and on the right, we have listed the norms embodied by these levels. One can see that a person can be perceived as behaving inappropriately or could be blamed because his or her behavior violates one of five norms that we have labelled harm, causality, avoidability, intentionality, and justification. For example, a person could be blamed simply because of the fact that he or she was associated (e. g., through temporal proximity or kinship) in some way with the actual perpetrator of harm, even though the person did not actually cause the harm. As

Table 8.1. The norms embodied by types of personal responsibility

I. The norms of proper conduct that people use when viewing harmful behavior.

Heider's (1958) Levels	Normative beliefs
1. Association:	harm should not occur
2. Causality:	one should not cause harm
3. Avoidability:	one should not cause harm carelessly
4. Intentionality:	one should not mean to cause harm
5. Justification:	one should not mean to cause harm with malevolent motives

II. Elaboration of three dimensions encompassed by these norms of proper conduct
Given that a perceiver has ascertained that an actor's behavior has resulted in harm, then that causality judgment may be further differentiated according to three dimensions:
1. *Avoidability* (unavoidability): the actor either could or could not have prevented the harmful consequence from occurring (e. g., because of factors like the actor's power, relative to environmental forces, or the actor's ability to foresee the harmful consequence).
2. *Intentionality* (absence of intentionality): the actor wanted and was trying to produce the harmful consequence (or the actor did not want and was not trying to produce the harmful consequence).
3. *Motive acceptability* (motive unacceptability): the actor's motives either were acceptable (i. e., harmful outcome was a means to achieve a further nonharmful end) or unacceptable (i. e., harmful outcome was either an end in itself or a means to achieve yet another socially disapproved or undesirable end).

another example, a person could be blamed because his behavior violates the intentionality norm irrespective of what his motives were. That is, a person can be blamed simply because he intended to cause harm.

In our view, the various issues regarding causality have to be considered before the perceiver can know whether any one of the five norms has been violated. Our assumption is that the perceiver asks whether the actor is linked to the harmful outcome by causality, avoidability, intentionality, and/or justification. That is, the perceiver must make several assessments, ranging all the way from simply establishing that harm has occurred to assessing whether the actor's motives for intending to cause harm were acceptable or unacceptable ones. We do not assume, however, that all of these considerations are made by all perceivers. Instead, the causal considerations made by a perceiver depend upon the norm or norms of proper conduct to which she or he subscribes in general, and also on which of these norms the perceiver believes is relevant to a particular harmful situation. Stated somewhat differently, the norm or norms to which a perceiver subscribes will determine whether he or she searches for particular types of causal evidence. For example, one perceiver may want to know only whether the actor's behavior was a sufficient condition for the occurrence of harm, whereas a second perceiver may want to know whether the actor could have avoided the harm. Of course, we need not assume that any of these judgments are made consciously (i. e., with awareness or deliberation) or that they are time-consuming to make. It is primarily when the harmful outcome is unexpected or extreme that the perceiver may try to make these judgments consciously.

In our view (Ferguson and Rule, 1983; Rule and Ferguson, in press), then, the norms of proper conduct or "oughts" to which a perceiver subscribes importantly affect his or her reactions to incidents, including harmful incidents. If the actor's behavior and/or manner of behaving are discrepant from (i. e., violate) an ought or oughts important to the perceiver, we believe that such an "is-ought" discrepancy (Kelley, 1973) can lead to the perception that the actor has behaved inappropriately. Perceiving the actor's behavior as inappropriate *can* then result in negative emotional reactions (such as anger), negative evaluative responses (such as blame), and negative behavioral responses (such as retaliation or punishment).

The notion that an is-ought discrepancy idea affects judgments of appropriateness is central in our analysis. There are, however, refinements to this idea. First, perceptions of inappropriateness and attendant emotional, evaluative, and behavioral reactions may be even stronger as the perceived is-ought discrepancy increases. Consider, for example, a perceiver who strongly believes that one should always behave as carefully as possible. If the perceiver assesses that the actor was quite careless, the perceiver may then view the actor's behavior as extremely inappropriate, compared to when the actor's behavior is only somewhat careless. Second, it might be suggested that perceivers (at least implicitly) rank order the various causality norms. For example, some perceivers may believe that it is worse to intend to cause harm than to cause harm because of carelessness. Third, perceivers may not only adhere to a hierarchy of causally-relevant norms, but also to a hierarchy of norms regarding the results of behavior per se.

For example, some perceivers may value a person's psychological or physical well-being more highly than a person's property. Given this evaluative hierarchy, personal injury would be viewed as more inappropriate than property damage. Finally, we suspect that the greater value given to certain behavioral results (such as personal injury) might lead to a shift in what causally relevant norms need to have been violated in order to lead to perceptions of inappropriateness.

In our adaptation of Heider's analysis of personal responsibility, several modifications have been made to Heider's ideas. Although Heider's levels have been viewed as representing a unidimensional continuum of personal causation, this idea has received little support in the empirical literature (e.g., Fincham and Jaspars, 1979, 1980; Harris, 1977). Such lack of support is consistent with Ferguson and Rule's suggestion that Heider's five types of personal responsibility can be better viewed as a series of nested causal dimensions (Ferguson and Rule, 1983; Rule and Ferguson, in press). The causal dimensions are causality, avoidability, intentionality, and motive acceptability. These dimensions are nested in the sense that one cannot have: (1) causality without some form of association, (2) avoidability without association of causality, (3) intentionality without association, causality or avoidability, or (4) motives without association, causality, avoidability, or intentions. Having recognized the nested nature of Heider's levels, we think it is important to reconceptualize this in other ways as seen in the lower panel of Table 1. We suggest that, given that the perpetrator is viewed as having caused the harm, the dimensions of avoidability, intentionality, and motive acceptability can be viewed as three independent dimensions, which can be factorially combined. So, for example, one can think of harmful incidents in which the harm, while being intended to achieve malevolent ends, was actually unavoidable. When viewed in this way, assessments can be made of whether the three dimensions are weighed differentially depending upon characteristics of the setting. Once again, note that we view these three independent dimensions (avoidability, intentionality, and motive acceptability) as normative ones. We also assume that the nonindependent dimensions of harm and causality are used normatively.

Finally, we have rejected others' views, based on Heider (1958), that blame is isomorphic with increasing degrees of personal causation. We propose instead that blame is dependent on the importance of the causally relevant norm to the perceiver and that assigned importance need not be isomorphic with the degree of personal causation embodied by a norm. Viewed in this way, one could say that these norms of proper conduct are more or less demanding regarding the nature of the evidence necessary to affirm or disconfirm their violation.

It can thus be seen that our analysis importantly differs from Heider's. These differences are not minor since they permit us to identify attributional considerations that may be made, to specify factors that may lead the perceiver to weigh differentially the available attributional information, and to delineate the interplay between attributional and normative considerations.

We have just emphasized the idea that harm violating a causally relevant norm or norms important to the perceiver can be responded to at least as if it were perceived as aggressive. At this time, we would like to summarize the evidence that bears on this proposition. Moreover, the research documents our position that negative reactions occur when there is an is-ought discrepancy and that these are stronger as the discrepancy increases.

Research with both children and adults shows considerable support for the idea that violation of causally relevant norms affects the perception of the appropriateness of harm. Although the discriminations made by children are not as great as those of adults (e.g., Ferguson and Rule, 1980; Fincham and Jaspars, 1979, 1980; Harris, 1977; Shaw and Sulzer, 1964), several studies have shown that blame, anger, and actual or desired retaliation are greater when the harmful event was foreseeable rather than unforeseeable, perpetrated intentionally rather than unintentionally, and perpetrated with unjustifiable rather than justifiable motives (e.g., Averill, 1979; Cohen, 1955; Dyck and Rule, 1978; Greenwell and Dengerink, 1973; Harvey and Enzle, 1978; Kulik and Brown, 1979; Nickel, 1974; Pastore, 1952; Weisfeld, 1972). For adults, at least, it in fact appears as though these considerations are more important than whether the harm has actually found its mark (e.g., Nickel, 1974). Note that this conclusion might have to be tempered for young children since some research shows that they are sensitive to outcome severity differences in their moral evaluations, as well as in their anger and revenge ratings (Denissen, 1982; Fergusen and Rule 1982; Rule and Duker, 1973). Tedeschi and his colleagues (Brown and Tedeschi, 1976; Tedeschi et al., 1974) moreover, have shown that retaliatory behavior that does not exceed the original provocation is seen neither as inappropriate nor aggressive compared to excessive retaliation, which indicates that perceptions of justice and fair play are very important in determining whether harm actually finding its mark is seen as aggressive. Another study of the Catholic University of Nijmegen demonstrated the same result with children as young as five years (Ferguson, Rule, and Wiegers, in preparation). Considered together, these studies show that violation of any *one* of the causally relevant norms affects the extent to which an actor's behavior is considered to be aggressive, as indexed by both involved and uninvolved perceivers' ascriptions of blame, reports of anger, and actual or desired retaliation.

Other studies (Averill, 1979; Denissen, 1982) go somewhat further than those just reviewed in that they could be construed as assessing how the importance of particular norm violations affects adult observers' reports of anger. For example, both adults and children report the greatest anger in response to unjustifiably intended harm, intermediate degrees of anger in response to foreseeably produced harm, and the least anger in response to justifiably intended and accidental harm. We interpret these results to mean that perceivers spanning a wide age range most highly value the justification and avoidability norms relative to causality and intentionality norms, which bears out the notion of a norm hierarchy. These results also bear out a point made elsewhere, which is that Heider's types of personal responsibility should not be viewed as a unidimensional continuum of moral culpability, but rather as a series of nested dimensions that can be assigned different weights depending on perceiver and event characteristics (Ferguson and Rule, 1980, 1983; Rule and Ferguson, in press).

The studies we have summarized show that the presence of an is-ought discrepancy influences how negatively an actor is perceived. Other research shows that increases in an is-ought discrepancy affect judgments of, and reactions to, the actor and his behavior (Harvey, 1981). For example, in our laboratory work at Nijmegen, we found that a retaliator was judged more and more harshly as the actor's behavior deviated more and more from what the children themselves thought he should do (Ferguson and Rule, in preparation).

Biasing Conditions

While our overview was brief, the results reveal in one sense a very rational view of adults' reactions to harmful incidents. It appears as though intent and inappropriate motives or actions are the conditions most likely to foster more extreme perceptions of inappropriateness. However, we wish to emphasize the possibility that the use of such criteria by perceivers represents the least biased case. Given the right setting conditions, we suspect that perceivers do not always respond in accordance with the relatively "rational" model. That is, violations of a less valued norm in one situation may give rise to just as strong a negative reaction as would violations of a more highly valued norm in another situation. While few studies explicitly document this notion, there is evidence consistent with it. For example, an angered person may not reduce retaliation upon receipt of information that there were mitigating circumstances for attack by the other person (Zillmann and Cantor, 1976). Or, when highly aroused by environmental stressors, such as exposure to heat, an individual may be motivated to harm a partner who is in no way responsible for the discomfort (Palamarek and Rule, 1979). From these results we can see that situational factors may make different norms salient to the perceiver. Simply the presence of aversive consequences or the fact that harm was done may be perceived as violation of a norm that is sufficiently important to preclude further normative considerations.

Research with adults provides only indirect evidence for our view that a variety of setting conditions influence which standards are used in responding to harm. Some of our own research with children goes somewhat further than research with adults in assessing the effects of setting conditions on which standards are applied. We have found, for example, that a child's role as the victim, transgressor, or an observer of harm systematically influences the norms that are applied in the situation (Denissen, 1982). Older children (10–13 years) especially use the norms of avoidability and motive acceptability more when role-playing the victim than when role-playing the observer or transgressor; the intent norm, however, is used the least by victims. Both transgressors and especially observers, on the other hand, respond as though stricter standards were being applied, in the sense that the actor has to have malevolently intended harm to elicit negative reactions. The norms children apply are also affected by whether the harm consists of property damage as opposed to personal injury and by whether the harm is mild or severe (Denissen, 1982; Ferguson, Olthof, Luiten, and Rule, in preparation; Ferguson and Rule, in preparation). While the effects are not identical across age groups, we consistently see that the actor is blamed more at lower levels of responsibility when the harm is either severe or personally injurious than when the harm is either mild or involves property damage (see also Elkind and Dabek, 1977). In addition, children make fewer discriminations according to the actor's personal responsibility when the harm is severe rather than mild or personally injurious rather than destructive. Such differences must reflect the role that values play in determining normative salience. That is to say, perceivers must differentially appraise the value of property over personal welfare and the value of "life over limb" (Pepitone, 1976). The more highly valued something is, the easier it is to annoy someone by threatening that value. All of the evidence we have just considered supports our proposition that, while a perceiver may be capable of applying any given norm to the analysis of harm, this does not guarantee that the norm will be applied. Instead, which norms are applied can depend on a variety of factors, including perceiver involvement and the nature of the harmful event.

In addition to setting conditions, expectancies about the harm-doer may also bias a variety of judgments. For example, research with both adults and children shows how important both category- and target-based expectancies can be in assignments of personal responsibility. We are all familiar, for example, with research showing how adults' racial stereotypes affect their labelling of an actor's harmful behavior (e. g., Duncan, 1976), and with research showing how adults' physical attractiveness or occupational stereotypes bias their memory for an actor's behavior (e. g., Cohen and Ebbesen, 1979) and can even cause them to behave in ways eliciting stereotype-consistent behaviors on the actor's part (e. g., Snyder, Tanke, and Berscheid, 1977). The biasing potential of adults' stereotypes has been shown even in situations where the actor's personal responsibility for harm was unambiguous (e. g., Nesdale, Rule, and Hill, 1978). While we know of no similar research with children, there is evidence that children's perceptions of how positively or negatively someone will behave differ depending upon the actor's sex, race, physical attractiveness, body build, and even according to the popularity of an actor's name (see Rule and Ferguson, in press). However, we suspect that the biasing potential of stereotypes is much weaker in children than it is in adults. Among other reasons, this suspicion is partly based on evidence that it is much easier to undermine a child's than an adult's impressions of someone based simply on changes in the valence of the person's behavior (see Rule and Ferguson, in press; Shantz, 1981). It appears that adults as compared with children, may use apparently more primitive bases for personal responsibility assignment, in part because of adults' greater sophistication in other domains (e. g., adults' greater ability to say that a characteristic is invariant across various transformations).

Other research verifies the importance of knowledge about the actor's characteristics in the personal responsibility assignments of both adults and children, although less of this research has focused on the characteristic of aggressiveness. For example, knowledge regarding a person's attitudes toward violence affects whether adults attribute malevolent intent to the actor for harmful behavior (Rosenfeld and Stephan, 1977). Our own research with children shows that even 5-year old children use knowledge about the person to predict the valence and causal nature of the actor's behavior (Ferguson, Olthof, Rule, and Luiten, in preparation; van Roozendaal, in preparation). However, children older than 8 years of age seem to rely more on such knowledge in making personal responsibility assignments for harmful behavior. From this perspective, then, we can see how much more biased older children and adults could be, relative to younger children, in how much their personal responsibility assignments are affected by the actor's past track record.

Conclusions

Our research, especially with children, reveals that many dimensions of causality may be used normatively, in the sense that blame is assigned, and anger and revenge are instigated. For example, lack of foreseeability in harming might engender negative reactions as strong as or stronger than malevolently intended harm. All of the results we have reviewed suggest that the perception of harm as malevolently intended (e. g., Mummendey et al., 1982); Tedeschi, et al., 1974) may not be a necessary condition for the labelling of an act as aggressive. At this stage, two possibilities can be entertained, however. Perceivers might shift in what they accept as evidence for inferring malevolent intent. Or, alternatively, as we have suggested, perceivers might use other standards to react as if an act were aggressive even though the perceiver may not view the act as malevolently intended.

A second major conclusion can be derived from our review. Our perspective and the results bearing on it have implications for an important assumption in the social psychological and the developmental literature on causality and sanctions. This assumption is that the application of norms embodying greater degrees of responsibility represents a shift from a primitive to a more sophisticated level of thinking. For example, it is implied in the literature that the use of the association criterion is more primitive than is the use of the causality criterion (e. g., Fishbein and Ajzen, 1973). We suspect that this may be true only for perceptions of causality, not for the application of norms. That is, we conclude that causality judgments may be cumulative and may develop cumulatively. While some authors (Fincham and Jaspars, 1979) have found no evidence for this idea when asking children and adults questions regarding how much the person caused the event, we have found evidence for the cumulativeness idea when using measures of perceived avoidability and intent.

However, despite the apparent cumulativeness of causality judgments, we do not find cumulativeness in the application of norms. For example, we have found that application of the norms of justification seems to develop earlier than the use of norms regarding avoidability or intentionality. Note that these data suggest a lack of isomorphism between increasing degrees of causality and blame. This supports our view that Heider's levels should be viewed neither as reflecting a series of developmental stages, nor as a unidimensional continuum of blame (Harvey and Rule, 1978). Instead, as a we have suggested both in terms of causality and norms, certain of Heider's levels and extensions thereof are better viewed as a series of independent dimensions, which can be weighed differentially.

Our study of children serves a function beyond that of examining age differences in cognitive abilities and normative appreciation. In particular, our research helps to validate the robustness of relevant attributional principles. In doing this, we have illustrated some invariances over age in the cognitive makeup of people. The differences observed in the data do not seem to reflect less basic cognitive abilities of children but rather their mastery of specific theories, scripts or other sources of knowledge. On the other hand, adults' failures to use various dimensions of causality under different setting conditions are not regressions to childish modes of inference. Instead, they seem to reflect strategies and priorities that have been learned over the life span, as a result of culturally shared ideas and exposure to uniquely distorted sources of data.

From a theoretical perspective, we have specified the link between attribution and aggression by focusing on blame ascriptions. We have also articulated how changes with age in social-cognitive skills and moral internalization may be implicated in these relations. Our analysis also extends current attributional perspectives to encompass more *social* attributional considerations, including norms, stereotyping, and ingroup-outgroup phenomena. Our analysis also indicates the many places at which the so-called rational attribution process may break down. That is, in contrast to past attributional analyses of aggression, we have taken into account how the hierarchy of norms subscribed to by a perceiver and group affiliations may affect assignments of causality, blame, anger, and the desire to retaliate.

General Issues

While our paper focused on how victims respond to harm-doing by another person, the focus and discussion of other papers raised some more general issues to which we will now turn. The construct of aggression was discussed from a variety of perspectives, including whether it is a social construct versus an objective quality of behavior or an individual versus an interactive concept. In the following section we will present our view about the aggression construct and its theoretical role.

The goal of theories of human behavior involving aggression as a construct is to provide statements or propositions that link antecedent conditions and their consequences. In doing this, some theorists have focused on propositions about eliciting cues and resulting harm derived from classical conditioning (Berkowitz, 1970), on propositions about how anticipated consequences affect instrumental harm doing (Bandura, 1973), or on propositions relating attributions, anger, and hostile aggression (Ferguson and Rule, 1983; Rule and Ferguson, in press). While these theories focus on cognitive, motivational and reinforcement processes of individuals, they are social psychological theories in the sense that norms (i.e., formal and informal rules of conduct) are considered and in the sense that the relations specify at least both an actor and a target (i.e., a dyad and/or a group). These theories then incorporate the social context and its meaning within their domains. What are the limitations of these approaches? Mummendey and her colleagues, Tedeschi, and Gergen (all this volume) – and various others – express dissatisfaction on a variety of issues. While we do not share all of their concerns, we have some that may overlap with some of those they are trying to express. In our view, the central limitation of our current social psychological theories of aggression rests on their static nature. They are essentially S-R or S-O-R theories that do not address the dynamic changes that occur over a sequence of events in time with corresponding changes in the state of the organism and the environment. Not only have the theories not addressed what stimulates an act of aggression at the outset, as Tedeschi (this volume) suggests, but they have not spoken to the changes that occur to the victim as a function of his response (and to the subsequent response by the actor). The iterative nature of the action sequence has been ignored. Specifying this process is the task ahead of us, regardless of whether we do so within our now more refined constructs and theories of aggression or within the newer uncharted area of coercive power. Although our focus was on the victims' anger and blame responses and on his aggression per se, we view the construct in the following way:

At a descriptive level, the term aggression can be defined as an act of commission or omission that actually or potentially involves harm to an individual, group, institution, or nation. The intentional production of harm can obviously serve a variety of motives or functions (including personal-instrumental, social-instrumental, expressive, or hostile ones). Harm need not necessarily be produced intentionally; it can be purely accidental or carelessly committed (Feshbach, 1964, 1971; Rule, 1974; Rule, Ferguson and Nesdale, 1979; Rule and Nesdale, 1976; Sears, 1961). Recognizing the avoidability, intentionality, and motive quality of harm sensitizes one to the fact that different types of aggression have different antecendents and/or consequences for all involved parties. This has been, and will continue to be, an important point for understanding under what circumstances harmful behavior will be expressed, as well as the nature of responses to it. Restricting one's definition of aggression to normatively inappropriate and intentionally injurious actions seems to us, therefore, to severely limit the scope of our daily encounters with harmful events. Such a restriction also ignores the possibility that there are stable individual and developmental differences in whether the same event is perceived similarly and thereby results in similar reactions or counter-reactions. We find it in fact paradoxical that so many discussions by the writers in this volume emphasized the subjective or social nature of aggression, while at the same time making arguments for a particular or restricted definition of aggression. In addition, such calls for definitional restrictions are not cogently based in theory. They derive instead from the fact that aggression defined as counternormative and intentionally injurious actions has empirically established predictive value. Yet, this should in no way detract from the potential predictive value of other types of harm, (Gergen; Tedeschi, both this volume). Moreover, while the victim's intepretation of harm as counternormative and intentional may be an important decision point in affecting the subsequent course of an interaction (Felson; Mummendey et al.; da Gloria, all this volume), such definitional restrictions seem to treat the victim as the crux of the interaction – leaving the transgressor in the position of only counter-arguing the victim's interpretation. This is somewhat surprising since several authors have reaffirmed the need to delimit the origins of aggressive incidents (Felson; Kornadt; Tedeschi, all this volume).

In the discussions in this volume, skepticism about the necessity of an anger construct has also been expressed. We view anger as an important theoretical construct because it provides the link between cognitions and behavior (Feshbach, 1964; Konečni, 1975). Without a motivational construct, cognitive theories fail to provide theoretical specification relating cognitions to behavior (see Bem, 1972; Kelley, 1971). Nonetheless, while anger figures prominently is explaining an individual's behavior, the antecedents of anger must be placed within the context of the dynamics of social interactions.

Finally, individual differences have not played a major role in our analyses, although many potential ones could affect perceptions of harm. For example, as we have already pointed out (Ferguson and Rule 1983; see also Kornadt; Zumkley, both this volume), an individual's level of moral development, authoritarianism, or high aggressiveness may affect responsibility levels assigned. Despite the potential relevance of these variables, no one has specified how, or whether, such individual differences modify the basic principles necessary to account for aggression. Consequently, we prefer to focus on understanding the basic processes of the phenomenon before turning to a study of individual differences. Moreover, measurement problems and the lack a of rationale for studying one individual difference variable rather than another lessens the impact of such research.

References

Averill, J. R. Anger. In H. Howe and R. Dienstbier (Eds.), *Nebraska symposium on motivation.* Lincoln: University of Nebraska Press, 1979.

Bandura, A. *Aggression: A social learning analysis.* Englewood Cliffs, N. J.: Prentice Hall, 1973.

Bem, D. J. Self-perception theory. In L. Berkowitz (Ed.), *Advances in experimental social psychology,* (Vol. 6). New York: Academic Press, 1972.

Berkowitz, L. *The contagion of violence: an S-R mediational analysis of some effects of observed aggression.* In W. J. Arnold and M. M. Page (Eds.) Nebraska symposium on motivation. Lincoln: University of Nebraska Press, 1970.

Brown, R. C., and Tedeschi, J. T. Determinants of perceived aggression. *Journal of Social Psychology,* 1976, 100, 77–87.

Cohen, A. R. Social norm, arbitrariness of frustration, and status of agent of frustration in the frustration-aggression hypothesis. *Journal of Abnormal and Social Psychology,* 1955, 51, 222–226.

Cohen, C. E., and Ebbesen, E. B. Observational goals and schema activation: A theoretical framework for behavior perception. *Journal of Experimental Social Psychology,* 1979, 15, 305–329.

da Gloria, J., and de Ridder, R. Aggression in dyadic interaction. *European Journal of Social Psychology,* 1977, 7, 189–219.

Denissen, K. "Kleine, Kleine stouterik; Zoudt gij moeder tergen?" The development of moral evaluation and the influence of socialization. Catholic University of Nijmegen: Unpublished Master's thesis, 1982.

Duncan, B. L. Differential social perception and attribution of intergroup violence: Testing the lower limits of stereotyping of blacks. *Journal of Personality and Social Psychology,* 1976, 34, 590–598.

Dyck, R. J., and Rule, B. G. Effect of retaliation of causal attributions concerning attack. *Journal of Personality and Social Psychology,* 1978, 36, 521–529.

Elkind, D., and Dabek, R. F. Personal injury and property damage in the moral judgment of children. *Child Development,* 1977, 48, 518–522.

Ferguson, T. J., Olthof, T., Luiten, A., and Rule, B. G. *Children's use of observed behavioral frequency vs. behavioral covariation in ascribing dispositions to others.* In preparation.

Ferguson, T. J. and Rule, B. G. Effects of inferential set, outcome severity, and basis for responsibility on children's evaluations of aggressive acts. *Developmental Psychology,* 1980, 16, 141–146.

Ferguson, T. J., and Rule, B. G. The influence of inferential set, outcome intent and outcome severity on children's moral judgments. *Developmental Psychology,* 1982, 18, 843–851.

Ferguson, T. J., and Rule, B. G. An attributional perspective on anger and aggression. In R. Geen and E. Donnerstein (Eds.), *Aggression: Theoretical and empirical reviews.* Academic Press, 1983.

Ferguson, T. J., and Rule, B. G. Age differences in moral evaluations of retaliation. In preparation.

Feshbach, S. The function of aggression and the regulation of aggressive drive. *Psychological Review,* 1964, 71, 257–272

Feshbach, S. Dynamics and morality of violence and aggression: Some psychological considerations. *American Psychologist,* 1971, 26, 281–292.

Fincham, F., and Jaspars, J. Attribution of responsibility to the self and other in children and adults. *Journal of Personality and Social Psychology,* 1979, 37, 1589–1602.

Fincham, F., and Jaspars, J. Attribution of responsibility: From man the scientist to man the lawyer. In L. Berkowitz (Ed.), *Advances in experimental social psychology,* (Vol. 13). New York: Academic Press, 1980.

Fishbein, M., and Ajzen, I. Attribution of responsibility: A theoretical note. *Journal of Experimental Social Psychology,* 1973, 9, 148–153.

Greenwell, J., and Dengerink, H. A. The role of perceived versus actual attack in human physical aggression. *Journal of Personality and Social Psychology,* 1973, 26, 66–71.

Harris, B. Developmental differences in the attribution of responsibility. *Development Psychology,* 1977, 13, 257–265.

Harvey, M. D. Outcome severity and knowledge of "ought": Effects on moral evaluations. *Personality and Social Psychology Bulletin,* 1981, 7, 459–466.

Harvey, M. D., and Enzle, M. E. Effects of retaliation latency and provocation level on judged blameworthiness for retaliatory aggression. *Personality and Social Psychology Bulletin,* 1978, 4, 579–582.

Harvey, M. D., and Rule, B. G. Moral evaluations and judgments of responsibility. *Personality and Social Psychology Bulletin,* 1978, 4, 583–589.

Heider, F. *The psychology of interpersonal relations.* New York: Wiley, 1958.

Hollingsworth, H. L. *Psychology and ethics: A study of the sense of obligation.* New York: Ronald, 1949.

Kelley, H. H. Moral evaluation. *American Psychologist,* 1971, 26, 293–306.

Kelley, H. H. The process of causal attribution. *American Psychologist,* 1973, 28, 107–128.

Konečni, V. J. Annoyance, type and duration of postannoyance activity and aggression: The "cathartic" effect. *Journal of Experimental Psychology: General,* 1975, 104, 76–102.

Kulik, J. A., and Brown, R. Frustration, attribution of blame and aggression. *Journal of Experimental Social Psychology,* 1979, 15, 183–194.

Mummendey, A., Bornewasser, M., Löschper, G., and Linneweber, V. It is always somebody else who is aggressive. A plea for a social psychological perspective in aggression research. *Zeitschrift fur Sozialpsychologie,* 1982, 13, 177–193.

Nesdale, A. R., Rule, B. G., and Hill, K. A. The effect of attraction on causal attributions and retaliation. *Personality and Social Psychology Bulletin,* 1978, 4, 231–234.

Nickel, T. W., The attribution of intention as a critical factor in the relation between frustration and aggression. *Journal of Personality,* 1974, 42, 482–492.

Palamarek, D. L., and Rule, B. G. The effects of ambient temperature and insult on the motivation to retaliate or escape. *Motivation and Emotion,* 1979, 3, 83–92.

Pastore, N. The role of arbitrariness in the frustration-aggression hypothesis. *Journal of Abnormal and Social Psychology,* 1952, 47, 728–731.

Pepitone, A. Social psychological perspectives on crime and punishment. *Journal of Social Issues,* 1976, *31,* 197–216.

Pepitone, A. The normative basis of aggression: Anger and punitiveness. *Recherches de Psychologie sociale,* 1981, 3, 3–17.

Rosenfeld, D., and Stephan, W. G. When discounting fails: An unexpected finding. *Memory and Cognition,* 1977, 5, 97–102.

Rule, B. G. The hostile and instrumental functions of human aggression. In J. de Wit and W. W. Hartup (Eds.), *Aggression: Origin and determinants.* The Hague: Mouton, 1974.

Rule, B. G., and Duker, P. The effects of intentions and consequences on children's evaluations of an aggressor. *Journal of Personality and Social Psychology,* 1973, 27, 184–189.

Rule, B. G., and Ferguson, T. J. Developmental issues in attribution, moral judgment and aggression. In R. M. Kaplan, V. S. Konečni, and R. W. Novaco (Eds.), *Aggression in children and youth.* The Hague: Martinus Nijhoff, 1984.

Rule, B. G., Ferguson, T. J., and Nesdale, A. R. Emotional arousal, anger and aggression: The misattribution issue. In P. Pliner, K. Blankstein, and T. Spigel (Eds.), *Advances in the study of communication and affect,* (Vol. 5). New York: Plenum, 1979.

Rule, B. G., and Nesdale, A. R. Moral judgments of aggressive behavior. In R. Geen and E. O'Neal (Eds.), *Perspectives on aggression.* New York: Academic Press, 1976.

Sears, R. R. Relation of early socialization experiences to aggression in early childhood. *Journal of Abnormal and Social Psychology,* 1961, 63, 466–492.

Shantz, C. U. Social cognition. In J. H. Flavell and E. M. Markman (Eds.), *Cognitive development,* a volume in: P. H. Mussen (Ed.), *Carmichael's manual of child psychology* (4th ed.) New York: Wiley, 1981.

Shaw, M. E., and Sulzer, J. L. An empirical test of Heider's levels in attribution of responsibility. *Journal of Abnormal and Social Psychology,* 1964, 69, 39–46.

Snyder, M., Tanke, E. D., and Berscheid, E. Social perception and interpersonal behavior: On the self-fulfilling nature of social stereotypes. *Journal of Personality and Social Psychology,* 1977, 35, 656–666.

Tedeschi, J. T., Smith, R. B. III, and Brown, R. C., Jr. A reinterpretation of research on aggression. *Psychological Bulletin,* 1974, 81, 540–562.

van Roozendaal, J. *The role of frequency vs. covariation in children's attributions for, and predictions of, prosocial and antisocial behavior.* Catholic University of Nijmegen: Master's thesis. In preparation.

Weisfeld, G. Violations of social norms as inducers of aggression. *International Journal of Group Tensions,* 1972, *2,* 53–70.

Zillmann, D., and Cantor, J. R. Effects of timing of information about mitigating circumstances on emotional responses to provocation and retaliatory behavior. *Journal of Experimental Social Psychology,* 1976, 12, 38–55.

Chapter 9

Social Justice and the Legitimation of Aggressive Behavior

Dieter Birnbacher

Introduction: The Contribution of Philosophy to Social Psychology

There are three ways in which philosophy, as I see it, is concerned with social psychological topics such as aggression:

1. Conceptual clarification. Philosophy brings to bear its analytical competence on the analysis of key concepts, which those working with these concepts in their everyday research often lack the necessary distance to view in a sufficiently detached and comprehensive perspective. Philosophical analysis should not be seen as a rival, but rather as a cooperative enterprise.
2. Rational criticism of methods used, interpretations imposed on the data, and possible normative proposals derived from these interpretations. In close cooperation with scientific methodology, philosophy contributes to the rational assessment of the credibility, import, and scope of a scientific theory in the light of such criteria as consistence, coherence, systematic unity, explicitness, rigor and evidential support.
3. The last and most representative concern of philosophy in respect to psychological research is the integration of psychological results in an overall view of human nature.

Whereas the work of most philosophers dealing with human aggression (and psychological theories of aggression) must be classified as falling into the third category distinguished, the remarks I shall offer will centre on the first task, the clarification and explication of central concepts. This will not always be simply an analysis of these concepts as they are actually used, either in everyday language or within the scientific community. Often, the explicatory enterprise will involve a restructuring of these concepts, insofar as their use is vague, incoherent, or obscure. It is no exaggeration to say that aggression is one of those psychological concepts for which such clarificatory work is most needed. Not only is there hardly any agreement about the exact content of the concept of aggression among the various researchers, there seem to be difficulties with this concept even in the individual theoretical approaches themselves. My thesis is that these difficulties are mainly due to the fact that the concept of aggression as used in scientific and nonscientific contexts is used partly in a value-*neutral,* purely descriptive way, and partly with a distinctly *normative* content, i. e., as an expression of moral *disapproval.*

Three Uses of the Concept "Aggression"

It is absolutely essential for understanding the role that moral principles, in particular principles of justice, play in the context of aggressive interactions to get clear about this distinction. The point is that this concept is not only used in different senses in different contexts, but that it operates on three completely different levels of discourse.

1. The first level is that of *conscious or unconscious motivation.* Aggression in this sense is a theoretical construct or explanatory device to account for various modes of behavior usually classed as "aggressive." Aggression is thought of on this level as a motivational source or reservoir from which aggressive acts flow, a potential that can be actualized, but need not be. That is, aggression in this sense need not be manifested in open behavior at all, since its expression can be inhibited or restricted to "sublimated" forms of expression: scientific controversy, artistic production, joke-cracking, sports, fast driving, and other kinds of socially accepted forms of behavior. It follows that this concept can be applied also to cases in which there is no aggressive interaction, but in which aggression is directed to oneself, e. g., constructively in asceticism, discipline, or self-mastery; destructively in self-contempt, depression, and suicide.

2. The second level is that of *open aggressive behavior.* Though it is difficult to give a strict definition of this concept in terms of necessary and sufficient conditions, it can be characterized by reference to paradigm cases, such as hurting others, attacking others, offending others verbally. One primarily thinks of aggression as a characteristic of individual acts or individual persons (in which case it is used as a disposition term), but obviously it can be applied as well to styles of interaction (war vs diplomacy, fighting vs law suits, etc.), to whole societies and cultures (the Kwakiutl vs the Hopi), as well as to biological kinds (wolf vs wombat). The important thing about this concept is that it is purely descriptive, i. e., value-neutral. As a consequence of this, it can be used in conjunction both with positive and negative valuations. Though, understandably, negative valuations of aggressive behavior are much more common, it is important not to overlook the fact that "aggressive," as a dispositional term, can also be used as a term of approval, e. g., with reference to hunting dogs, sports champions, sales managers, soldiers, revolutionaries, and potential sex partners. In psychotherapy certain forms of open aggression are often encouraged in order to counteract such tendencies as submissiveness and self-aggression.

3. The third use of the term "aggression" and its cognates is the *inherently pejorative* one. In this sense, "aggression" is used in condemnation, moral or otherwise, of an act, an intention, or want (or the disposition to perform the act or to

have the intention or the want). Obviously, there is not only one, but indeed a great many, variants of this concept, differing in the descriptive elements combined with the negative valuation.

Principles of Social Justice and Their Role in Aggressive Interaction

We are now in a position to indicate how principles of social justice are involved in the legitimation of aggressive interactions:

1. They are, trivially, used in justifying the condemnation of an aggressive act ("aggressive" understood in the second sense), that is held to be unjust, contrary to justice, unfair, unjustified, etc., or, alternatively, in justifying the use of the expression "aggressive" in its disapproving sense.
2. They are invoked, explicitly or implicitly, by the "aggressor" himself, or others, in justifying (or excusing) some piece of aggressive behavior.
3. Finally, they are invoked in justifying corrective, counteracting, or retaliatory measures others may take in response to an aggressive act directed at themselves or at a third party.

The third case is perhaps the most interesting one and surely the most complicated to deal with. The complication results from the fact that in this case the act of justifying the counteracting move may in itself constitute a step in an aggressive exchange, e.g., in political debate or in war propaganda. The justification, rationalization, and ideological underpinning of the counteraggressive move (or the attempt at such a justification) may itself constitute, or be looked upon by the other party, as an act of aggression over and above the aggressive content the countermove may have by itself. One might talk in such cases, paradoxically though it seems, of *moral aggression*. By putting the opponent in a morally disadvantageous position, either in the perspective of others or his own perspective, aims similar to those achieved by other kinds of aggression are secured. It is a well-known fact of clinical psychology that moral pressure can be a very effective means of coercing others' behavior, e.g., in what is sometimes called "depressive blackmailing," possibly involving suicide threats. Moral aggression can also be an important element in sadomasochistic interactions in which the "sadism" of the "masochistic" party consists precisely in imposing guilt feelings on the "sadistic" partner.

Moral Versus Nonmoral Reasons

Evidently, not every reason one might be tempted to put forward in defense of an aggressive act is a *moral* reason. Saying, e.g., "I just feel provoked by your behavior towards me," if it is a reason at all, is not the kind of reason one might want to call a moral reason. No reason is a moral reason that judges an action from a purely individualistic point of view, in terms of individual desires, interests, or predilections. The "moral point of view" is an inherently *general* one, from which one looks at the action, the action context, and the actor as instances of certain *types* of action, action context, and actor. One looks at the concrete, individual action as an instance to which some universal rule applies, from a detached, "objective" viewpoint. In the famous words of Hume (1963, p. 272):

"When a man denominates another his *enemy*, his *rival*, his *antagonist*, his *adversary*, he is understood to speak the language of self-love, and to express sentiments, peculiar to himself, and arising from his particular circumstances and situation. But when he bestows on any man the epithets of *vicious* or *odious* or *depraved*, he then speaks another language, and expresses sentiments, in which he expects all his audience are to concur with him. He must here, therefore, depart from his private and particular situation, and must choose a point of view, common to him with others! He must move some universal principle of the human frame, and touch a string to which all mankind have an accord and symphony."

Hume is here pointing to the 2 defining characteristics of moral justification.
1. *Moral judgments must be derivable from moral principles* prescribing or permitting an action of a certain type under circumstances of a certain type for an agent of a certain type. This is known as the "principle of universalizability". Moral judgments must be derivable, that is, from principles that do not contain proper names for agents, actions, or situations, or such quasi proper names as "I," "this", or "here." It follows that situations similar in all relevant respects must, if they are to be judged morally, be judged similarly. Likewise, there can be no moral judgment that privileges just *me* over and above others unless I am in a position to invoke some privileging characteristic I myself possess, while others don't, and which I honestly think to be morally relevant.
2. *Moral judgments claim universal assent.* Moral judgments are addressed to a faculty which is shared by all men and yields the same judgment if properly used. Rationalists (e. g., Kant) identify this faculty with practical reason, moral insight, or value intuition; emotivists (e. g., Hume) with value feeling, natural sentiment, sympathy, or the like.

These are the standard conditions for a justification to count as a moral justification, and it is easily seen that the restrictions imposed by these conditions are rather weak. First, the universal principles from which moral judgments are supposed to follow can be highly specific. The restrictions imposed by the standard conditions do not rule out that these principles differentiate between moral duties or moral rights according to such features as sex, race, status, or social class. Second, the conditions do not exclude from the domain of moral principles any version of universal *egoism*, understood as the position that everyone is morally permitted to use all means available to him to further his own interests. Variants of this position are better known under the name of *social Darwinism* and *nationalism*. (It is clear that this position is favored mainly by those who expect to gain, and not to lose, by its general acceptance.) If universal egoism is identified with rational egoism (as those versions propounded by philosophers usually are), it justifies the use of only those means that can reasonably be expected to actually promote the interests of the party concerned and allows the use of aggression only insofar as it can reasonably be expected to pay for the aggressor in the long run. It does nothing, then, to legitimize aggression *for its own sake* – unless, improbably, the desire for aggression itself is so dominant that everything else, including the consequences of the aggressive act, are a matter of indifference to the agent. Thus, in a society in which everyone is a universal egoist, there will not only be a perfect consensus about the fundamental norms governing social interaction (if there is conflict, it is not about the rules of the game, but is part of what playing the game consists in), there will also be a high chance of having fixed expectations about how the other members of the society will act. Naturally, this advantage will only apply if there is certainty that all the members are universal egoists, which is problematic enough, since not every rational egoist, if he is really rational, will let the others know that he is an egoist. Perhaps only the very strong and mighty can afford to be so frank about their principles.

Procedural Versus Substantive Justice

What I want to claim, now, is that insofar as justification of aggression rightly claims to count as *moral* justification, principles of *justice* are of particular importance, especially in the case of *collective* aggression. Nearly all revolutions, and most wars, have been initiated in the name of justice, the adversary usually invoking the same principle on his part in fighting back. Strikes are usually justified by principles of social justice (getting one's just share), but so are eventual countermeasures taken by the employers. This is not at all surprising. Among moral principles, principles of justice are commonly held to have a particular force moral principles otherwise lack. "Justice is a name for certain moral requirements, which regarded collectively, stand higher in the scale of social utility, and are therefore of more paramount obligation, than any others." (Mill, 1964, p. 59). That explains why they are so frequently appealed to in justifying acts of aggression, which, from their very nature, stand very heavily in need of justification. At the same time, justice, by itself, is one of the vaguest and most polymorphous concepts, a *Leerformel* (Topitsch, 1960) lending itself to being invoked by the most diverse parties in support of their own positions. What is clear about the concept is that it has positive *normative* content, that it is used to describe some state, procedure, or institution as normatively satisfactory; what is less clear is that it has any definite and fixed *descriptive* content. We cannot naively assume that there is some one thing called "justice" that can be found or discovered if one goes on looking for it carefully enough, something like the platonic form or "idea" of justice representing the essence of justice over and above the welter of varying conceptions and interpretations of it. There simply is no such "idea," open to inspection by everyone, representing the "true" essence of justice. It is simply not feasible to come to terms with relativism as easily as that. On the contrary, there is hardly any concept to which the platonic picture is less adequately applicable.

The concept of justice is systematically ambiguous, and to clear up this ambiguity it is important first to distinguish between what has been called *procedural justice* and *substantive justice*. Theories of procedural justice apply the concept to certain procedures, such as majority voting, contracts arrived at by fair bargaining, or free play of the market forces, irrespective of the results to which these procedures may lead. Theories of substantive justice, on the other hand, allow an estimate of the result itself as more or less just, independently of the procedure by which it has been reached. Obviously, norms of procedural justice are usually proposed with a view to the results the procedure is expected to yield, these results having antecedently been judged to be desirable on other grounds (often, but not necessarily, grounds of substantive justice). Thus, the procedure called fair trial is usually justified not only by its inherent justice or fairness, but also by the probability that it leads to results (the punishment of the guilty and the nonpunishment of the nonguilty) that are themselves just in a substantive sense. In other cases, it is less clear that the justice of the results of the procedure are generally as just as the procedure itself. Though the market system – the primary aim of which is efficiency, rather than justice – may be rightly held to constitute a just procedure, a salary or price or distribution of incomes or market power to which it leads may simultaneously, and with perfect consistence, be looked upon as highly unjust on substantive principles of justice. In other cases, again, it is doubtful whether the results can generally be judged in terms of justice at all, such as in parliamentary legislating.

There are only few theories of justice that attempt to limit the concept of justice to procedural justice (Nozick's [1974] recent theory belongs to this group), or attempt to derive from a unified procedural account the various substantive concepts commonly applied in contexts in which a purely procedural concept cannot be used (this is the approach of Rawls' [1971] theory). In Rawls' theory, which reserves the name "fairness" for the procedural concept of justice, the justice of a decision, law, institution, etc. is equated with the result of a *fair* decision procedure. This is defined as operating in a hypothetical state of perfect equality behind a "veil of perception" which hides from those making the decision their actual social positions, thus preventing them from choosing principles of justice which favor the positions they actually hold in society. The veil of perception creates an artificial Humean objectivity, insofar as none of those deciding on the principles of justice is in a position to privilege his own preferences. This ideal version of a social contract theory of justice is a purely a priori one, and it has to be in order to give a minimum of plausibility to Rawls' claim that he actually knows what the principles are that would be chosen in his hypothetical "original state."

It goes without saying that principles of procedural justice are especially important in a pluralistic society with widely divergent interests, values, and ideologies, including divergent conceptions of substantive justice, if conflicts are to be resolved in a nonaggressive, nonviolent manner. There may be unanimity about a decision procedure in spite of harshly conflicting opinions about the rightness or justice of individual decisions. Often, however, principles of procedural justice themselves stand in conflict with one another, thus motivating acts of collective aggression even within society (revolutionary upheavals, civil wars, guerilla fighting). The aggressive interactions between groups representing democratic ideals and groups representing autocratic or oligarchic ideals (as those in the communist satellite states during the last decades), or between basis-democratic or anarchistic minorities and a majority representing parliamentary democracy (as in the various forms of left-wing violence in the Western democracies after 1968), were not only motivated but also explicitly justified by principles of procedural justice irreconcilable with those enforced by the powers that be. Sometimes, however, the principles of procedural justice invoked in such conflicts can be interpreted more plausibly as being only secondary to principles of substantive justice, such as principles of just distribution of income, believed to be more easily realizable on the basis of changed procedures. Even more widespread than aggressive conflicts between groups maintaining conflicting principles of procedural justice are aggressive conflicts between groups, one of which defends as "just" a procedure, A, by which a decision, B, has been reached, whereas the other appeals to a principle of substantive justice on which decision B is manifestly unjust. Interestingly, this kind of conflict may occur even if the first group is no less convinced of the injustice (in a substantive sense) of B than the second group, as, e. g., in the controversies about blocks of appartments left vacant from motives of financial speculation. If the first group puts heavy weight on procedural principles, such as the rule of law and the protection of private property, it is likely to provoke aggression on the part of protest groups regarding the (substantive) injustice involved in the use of the property as sufficient moral justification for seeing their act of breaking the law (in occupying the houses) as an act of legitimate civil disobedience.

The distinction made between procedural and substantive justice is no more than a first step in the process of disentangling the systematic ambiguity of the concept of justice. For much more frequent than conflicts between procedural justice and substantive justice are conflicts between mutually incompatible principles of substantive justice.

Some Further Distinctions

Perhaps the most fundamental principle of substantive justice is: Treat alikes alike. This principle, sometimes called "formal justice," explains the close relation that has traditionally been construed between *justice* and *equality*. It is often said that an unjust law, e. g., one that prescribes disproportionately grave punishments for minor offenses, would even be more unjust if it allowed for arbitrary exceptions. I am not sure whether this holds unconditionally, since there in fact are laws so brutal that anarchy would presumably be preferable. Nonetheless, it is clear that arbitrarily unequal treatment of equal cases constitutes an element of injustice over and above the injustice of the treatment itself. Obviously, the principle of "formal justice" is nothing but the application, to principles of justice, of the principle of universalizability holding for all moral principles. It follows from the Humean characterization of the moral point of view as an impartial and objective one. Variant readings of this central, metaethical principle are such popular and often quoted moral principles as the Golden Rule or Kant's categorical imperative. Both require moral evaluations to be universalizable, i. e., to be derivable from principles applying to a whole class of persons, situations, and actions.

The material principles of substantive justice that remain to be considered can again be subdivided into two major classes termed (following Aristotle) *corrective* and *distributive* justice. Aristotle showed that both have something to do with "right proportions," though they operate in completely different kinds of situations. *Corrective* justice operates in "voluntary transactions," such as selling, buying, lending, and depositing, as well as in "involuntary transactions," such as theft, adultery, and assassination. Problems of corrective justice are, then, problems about which price, which compensation, which salary is just in the sense of standing in the right proportion to what they are the compensation for (voluntary transactions), or what is the appropriate punishment or the appropriate damages in the case of involuntary transactions. The situations in which problems of corrective justice arise are typically those in which services or disservices, goods or bads of some kind, are exchanged between parties roughly on the same level.

Problems of *distributive* justice, on the other hand, arise whenever "honour, wealth and other divisible assets of the community are allotted among its members in equal or unequal shares" (Aristotle). Obviously, problems of distributive justice arise not only in the distribution of public goods, such as public services, funds, or opportunities, but also in the distribution of public "bads," such as duties, taxes, and fees for public services. In all these cases there is no exchange of goods or bads, but a distribution of benefits or burdens by an agency on a superior level to those among whom the distribution is effected (state vs citizens, parents vs children, teacher vs pupils).

I shall mention only two principles of corrective justice much discussed in moral philosophy: fairness and retribution. The interest, from a theoretical point of view, of these principles lies mainly in the fact that, at least prima facie, they seem to be incompatible with *utilitarian* modes of thought, which are predominant in present-day normative ethics (Hoerster, 1971). That means, if these principles of corrective justice are accepted, *justice* would be opposed to *moral rightness* defined on utilitarian lines. *Fairness* is the principle that everyone who reaps benefits from a joint enterprise should contribute a corresponding share of the costs of the enterprise. [This principle can with equal right be interpreted as a principle of distributive justice, i.e., as a principle of the *distribution* of benefits and burdens, as it is by Lyons (1965, p. 164).] *Retribution* is the analogue of fairness in the sphere of punishment. Like fairness, retribution can be understood both as an answer to the question, who is to be punished (who is to pay), and to the question, how is the guilty to be punished (how much shall he pay). The general principle of retribution, that only the guilty are to be punished and that nobody who is not guilty must be punished, is an answer to the first question. This retributive principle is frequently, and understandably, advocated even by those who do not think that retribution is the right principle to assign kinds and degrees of punishment to offenses. However, even if retribution is accepted as an answer to the second question, too, it is still open to different interpretations. What retribution means depends, on the one hand, on what is taken as the basis of retaliation, the gravity of the act taken in itself, or in terms of consequences (what harm has been done?), the degree of guilt with which it has been done (are there excusing or mitigating conditions?), or the moral quality of the character traits of the agent from which it has flown. A concrete standard of retaliation, on the other hand, must specify a measure by which kinds and degrees of punishment are apportioned to degrees of gravity of the act to be punished. It is clear that acceptance of an essentially retributive theory of criminal punishment does not necessarily imply such drastic and disproportionate interpretations of the *lex talionis* as are as presently practiced in some Arabian countries. Adherence to the principle of retribution requires no more than that more serious offenses are invariably punished more severely than less serious ones, however "severity" is understood in concrete terms.

Conflicting Models of Just Distribution

Principles of corrective justice can, in turn, conflict with principles of distributive justice, but more important in the context of aggressive conflicts are divergences between various conceptions of distributive justice themselves, leading to conflicting claims on the distributive agency, such as the state. The mutual incompatibilities between the principles of distributive justice are notorious. They will become apparent from a short enumeration of these principles which essentially follows the classification given by Perelman (1945).

1. *Egalitarianism of Treatment.* This is the conception that everyone should receive the same treatment, irrespective of individual differences. This principle usually underlies the allotment of protection of civil rights provided by the legal system. It is also applied, mostly for efficiency reasons, for prices, fees, and indirect taxes.
2. *Egalitarianism of Results.* This conception requires differential treatment with a view to securing, or contributing to, an equality of results. It underlies practices such as transfer payments from taxes, public insurance schemes, family allowances, compensatory education, and the like. It overlaps with conception 6 below, differential treatment according to needs.
3. *Differential Treatment According to Merit.* This conception is operative whenever praise or blame are differentiated according to the moral quality of an act. In other cases merit is equated with nonmoral qualities like skill or achievement, as in the distribution of wages, examination degrees, and prices.
4. *Differential Treatment According to Effort.* This conception underlies part of the teacher's practice of praising and blaming pupils, or the criteria of good conduct used in shortening prison penalties. Apart from special cases as these, it is rarely applied because it encourages the untalented. Intuitively, however, it certainly is one of the most convincing principles of distributive justice.
5. *Differential Treatment According to Rank.* This conception is largely outdated but still operative in the distribution of honors, in the remuneration of politicians, in matters of protocol, etc.

6. *Differential Treatment According to Needs.* This principle has found its most famous expression in Marx' critique of the Gotha program of 1875: "From each according to his abilities; to each according to his needs" (Marx 1971, p. 339). This principle is superficially similar to the compensatory principle 2 but is nevertheless different, in that it does not refer to the (probable) results of a distributional decision, but to the needs present at the very moment of the decision (i.e., to the relative strength of the needs as felt at the period of the decision instead of the eventual levelling effect it may produce). There is, of course, again a great deal of ambiguity in the concept of "need" itself, which can refer to needs as actually felt, to needs as professed in responses to questions, or to needs recognized by society as worthy of satisfaction.

Though all these conceptions of distributive justice are mutually incompatible, there need not necessarily arise a conflict because sometimes their scope is severely restricted. For example, the scope of the principle of distribution by effort is so narrowly circumscribed that it does not generally conflict with the principle of distribution by merit. Conflicts, and the need for compromises, arise only with principles of larger scope, such as the principle of distribution by merit and the principle of distribution by needs, as in penal law, social services policy, tax policy, wage policy, and health care. Historically, in the predominant conceptions of social justice, a major shift has occurred during the last century. Whereas, in the 18th and 19th centuries, the primary political conflict was between representatives of distributive justice based on rank and representatives of distributive justice based on merit, in the 20th century political conflicts are primarily about the relative weight of various principles of distribution by merit, on the one hand, and various principles of distribution by needs, on the other. The evolution of political ideas is still going on, but its general tendency is unmistakably toward the principle of distribution by needs and the principle of equality of results, as exhibited in compensatory education and extensive social transfer payments, and the replacement of retributive ideas, in the practice of punishment, by ideas of individualized treatment.

The fact that in a pluralistic society there coexist incompatible, though overlapping, principles of distributive justice gives rise to potentially aggressive conflicts in two ways.

First, A and B hold different conceptions of social justice such that, according to A's principles, A should get a larger share of some centrally distributed good than he has, whereas, according to B's principles, B should get a larger share than he has. That is, A tends to consider any distributive policy unjust that will not provide him the larger share, which he thinks is due to him. So thinks B in respect to himself. Therefore, clashes between A's and B's claims can be avoided only by a policy that either makes the cake bigger so that both A and B can be satisfied or by changing A's or B's ideas of social justice. Since there are goods, the quantity of which cannot possibly be increased, such as power, status, or relative position in the social hierarchy, clashes are to be expected whenever the second policy does not work. A more complex variant is the possibility that, according to A's principles, not only should A get a larger share than he has, but B should get a smaller share than he has, whereas, according to B's principles, it is the other way round. In this case, obviously, no increase, however great, of the cake can satisfy both A and B.

Second, A and B hold different conceptions of social justice such that, according to A's principles, C should get a larger share than he has, whereas, according to B's principles, C deserves nothing better, or even less. (This is the situation of the bourgeois communist revolutionary or any social reformer with "idealistic" motives.) In both cases, the good judged to be unevenly or unjustly distributed can belong to any one of the three central dimensions of social inequality: economic position, status, and power.

It is characteristic for a society with a plurality of coexisting principles of justice that any group which has certain distributive interests in common has quite a good chance to appeal to considerations of justice to have its interests recognized by society at large. This is partly due to the fact that the positive ring of the concept of justice works independently of the specific content associated with it. Partly, it is a consequence of the fact that in a pluralistic society there is a certain preparedness to listen to arguments from justice, even when they are ultimately unacceptable from one's own point of view. For there is hardly anybody with allegiance to only *one* conception of justice. On the one hand, one will at least hold different conceptions of distributive justice for different contexts of application. The socialist with egalitarian ideas concerning income may well be a strict nonegalitarian concerning power; as a matter of fact, he has to be, if he wants to be consistent, since it does not need much political imagination to figure out that the maintenance of a strictly egalitarian distribution of income requires a firm centralization of power. Analogously, accepting an egalitarian position concerning individual chances requires accepting large inequalities in resultant income, since the talented will gain social premiums the untalented will have to forgo. On the other hand, one's conceptions of distributive justice commonly are a combination or mix of the principles listed above. Purism in matters of justice has become rare. Even those who put great weight on the principle of distribution by merit will not be wholly opposed to social insurance schemes working on the line of the principle of equality of results. And even the socialist with strong egalitarian tendencies will to some extent accept unequal wages from incentive considerations. Our ideological reality is characterized not by any one principle of justice, but there are all kinds of compromises, midway solutions, and integrating models. The English metaphysician F. H. Bradley once said that metaphysics is the finding of bad reasons for what we believe upon instinct (Bradley 1930, p. X). With only slight exaggeration it might be said that politics has largely become the business of finding bad justifications, in terms of justice, for what one wants from interest. The reason why appealing to principles of justice has become so popular in the rhetoric of trade unionists and professional lobbyists is that in our society there is a good a priori chance for the most diverse criteria of justice to find an audience, at least temporarily.

The Utilitarian Way Out

The philosopher can react to this rather desperate situation either with profound scepticism as to whether the plurality of principles of justice can in some way be reduced (this is the view of Perelman), or he can declare some one mixture of the various conceptions of justice (fixed by trade-off relations among the principles) to be the *right* one (this is Rescher's [1966] solution), or he can try to find some common denominator by which the diverse conceptions of justice can be made commensurable and be subjected to some kind of comparative assessment (this is Mill's strategy in *Utilitarianism* [1964]). The first reaction is sheer resignation. The second is unsatisfactory because of the high degree of relativity of the trade-offs between the criteria of justice different people find convincing. The only promising way to come to terms with the matter seems to me to be the utilitarian approach.

The utilitarian approach consists essentially in leaving principles of justice to themselves and going back to the *interests* underlying these principles. By choosing this procedure, the utilitarian does not want to imply that there are no autonomous *moral* interests, such as a sense of justice or genuine conscientiousness. He is not bound, by the perspective he assumes, to look upon justice as nothing but an ideological clothing of personal interests, as a truly reductionist perspective, on the lines of Nietzsche, Marx, or Freud, does. Nevertheless, the utilitarian knows as well as everyone that the motives behind the acceptance of principles of justice are quite often of a nonmoral sort: principles of retribution can be motivated by revengefulness, by motives of self-protection, or by a desire to see one's anger and one's hurt feelings publicly expressed; principles of distributive justice by envy (if they are advantageous to one's own and disadvantageous to someone else's position) or guilt feelings (if they are advantageous for someone who is noticeably worse off than oneself). Though aware of these psychological data, the utilitarian will eschew any depth-psychological reductionism. He will not go so far as to declare that those acting from a sense of justice are "really" serving some personal psychological or economic interest. His perspective is, rather, a functionalist one. He will look at principles of justice as social objectivizations of very fundamental human interests, as well as of the emotions and motivations connected with them. In this way, he will not only be able to render these principles intelligible in terms of a common denominator (human interest), thus bridging the gaps between the principles, but also to assess the role played by them in social interaction, thus opening a possibility of evaluating alternative conceptions of justice from the vantage point of overall social utility. In the case of a conception of justice, this utility comprises both the utility of *acting in accordance with* the principle, as well as of the attitudes and expectations that go together with the *acceptance* of the principle, and the general orientation of thought implied by judging others in accordance with it and passing it on to the next generation.

I shall give two examples in order to illustrate how such a functionalist reductionist perspective might work out in the context of principles of justice. The first one concerns the conception of distributive justice as distribution according to merit or achievement. This conception can simply be understood as the moral objectivization of an incentive/deterrence system encouraging achievement and behavior in conformity with society's moral norms. The concept of merit, together with its psychological manifestations in attitudes, emotions, and desires, can then be seen as essentially a social device for the allocation of human resources in the direction of socially desirable ends. Looked at in this way, the apparent incompatibility between the backward-looking principle of distribution by merit and strictly forward-looking utilitarian principles is seen to dissolve, paving the way for a reinterpretation of the distributive principle as, at least in part, a secondary principle in the utilitarian's sense, a principle translating the primary principle of social utility maximization into concrete terms.

My second example is that of retribution. Retribution can be looked upon as a paradigm of a moral concept that cannot be fully understood without recourse to the sphere of interests and emotions lying behind it. This has recently been very convincingly shown by Mackie (1982). The models proposed by various philosophers, among them Kant and Hegel, to justify retributive punishment either beg the question or fail to demonstrate what they purport to demonstrate. If punishment is intended to make the offender *pay back his debt* to society, this intention is misconceived in the majority of cases because, in general, society does not gain any advantage from the punishment of the offender (apart from the satisfaction of its vindictiveness). If, again, it is said that retribution *annuls* the crime, this is no less incoherent, since the crime cannot be annulled in the way the criminal can. If it is said that retribution *restores a fair balance* between criminal and society, this model is again inadequate. For what can be restored, once the offense (or the act of aggression) has been committed, is the balance of advantages and disadvantages both parties have gained from the act. The model explains why we expect the criminal to compensate the damages caused by his act, but it does nothing to explain why we should punish him in a retributive way. From these and other considerations, Mackie draws the conclusion that in order to understand retributive practices and retributive emotions we should rather look to the biological and social *functions* of retributive norms. As soon as we give up looking for an abstract justification of retributive norms in the ethical sphere and start to ask questions about the *point* these norms have in the functioning of society, it comes to dawn on us that retribution originates as a device of discouraging the aggressor from repeating his attack, what in fact it still is, on a social, as well as on a biological, level. For individuals and whole species alike, retaliation has an evident survival value. In this way, the development of retributive (and inherently aggressive) emotions, such as resentment and vindictiveness, becomes intelligible as supporting, on the psychological level, retaliatory behavioral tendencies. These emotions summon the energies required for effective retaliation and motivate social cooperation in collective revenge. The rise of abstract principles of retributive justice is the last, most "sublimated" stage in this development. Retributive norms of justice objectivize and rationalize the retributive emotions, making retribution independent of the actual occurrence of these emotions. In the modern legal system, finally, the work of retribution is done largely without any disturbing interference from the original retributive emotions.

Conclusion

The interconnections between justice and aggression are complex and involved. There is, however, also a very simple one to which I would like to draw attention in conclusion. We know that there is no simple recipe for how to keep aggression at a minimum in a society. But at least one of the preconditions of a low degree of aggressiveness is that norms of procedural justice are firmly entrenched. The very idea of civilization is bound up with the process of inventing and enforcing nonaggressive procedures for resolving aggressive conflicts, thus diminishing the self-destructive potential of aggressive interaction. Not unlike their biological analogues, the ritualized forms of aggressive interactions of animal species, which by unimpeded intraspecific aggression would jeopardize the survival of the species as a whole, principles of procedural justice are indispensable for preserving the inner peace of a society by channeling aggressive conflicts between opposed conceptions of substantive justice. Barry (1965, p. 105) has put the point succinctly: "The more society is divided on substantive values the more precious as a means of preserving social peace is any agreement that can be reached on procedure."

References

Aristotle. *The Nicomachean ethics*. London: Heinemann, 1934, bk. V, ch. 2, 1130 b 31.

Bradley, F. H. *Appearance and Reality*. Oxford: Clarendon Press, 1930.

Barry, B. *Political argument*. London: Routledge & Kegan Paul, 1965.

Hoerster, N. *Utilitaristische Ethik und Verallgemeinerung*. Freiburg/München: Alber, 1971.

Hume, D. Enquiry concerning the principles of morals, sect. IX, pt. 1. In: D. Hume, *Enquiries* (Ed. L. A. Selby-Bigge). Oxford: Clarendon, 1963.

Lyons, D. *Forms and limits of utilitarianism*. Oxford: Clarendon, 1965.

Marx, K. Kritik des Gothaer Programms. In: K. Marx und F. Engels, *Ausgewählte Werke*. Moskau: Progress, 1971.

Mackie, J. L. Morality and the retributive emotions. *Criminal Justice Ethics*, 1, 1982. 3–10.

Mill, J. S. Utilitarianism, ch. 5. In: J. S. Mill, *Utilitarianism, liberty, representative government*. London: Everyman, 1964.

Nozick, R. *Anarchy, state, and utopia*. New York: Basic, 1974.

Perelman, C. *De la justice*. Bruxelles: Presses Universitaires, 1945.

Rawls, J. *A theory of justice*. Cambridge (Mass.): Harvard University Press, 1971.

Rescher, N. *Distributive justice. A constructive critique of the 'utilitarian theory of distribution.'* Indianapolis/New York: Bobbs-Merrill, 1966.

Topitsch, E. Über Leerformeln. Zur Pragmatik des Sprachgebrauchs in Philosophie und politischer Theorie. In: E. Topitsch (Ed.), *Probleme der Wissenschaftstheorie. Festschrift für Victor Kraft*. Wien: Springer, 1960.

Author Index

Ajzen, J. 144, 150
Anscomb, G. E. M. 7
Arendt, H. 70
Athens, L. H. 108
Atkinson, J. W. 23
Austin, J. L. 51, 53, 58
Averill, J. R. 147

Bandura, A. 9, 10, 17, 55, 151
Barker, R. 84, 85
Baron, R. A. 55, 85
Barry, B. 170
Bartoszyk, G. D. 91
Beaman, A. L. 17
Bem, D. J. 153
Berkowitz, L. 17, 25, 36, 55, 70, 77, 101, 136, 151
Bernstein, F. 70
Berscheid, E. 149
Billig, M. 101
Birdwhistle, R. J. 11
Black, D. 109, 110
Blakar, R. M. 57
Bonoma, J. V. 13, 18
Bornewasser, M. 27, 66, 76, 77, 79, 80, 86, 88, 89, 92, 99, 101, 102, 143, 150
Bouffard, D. L. 86
Bradley, F. H. 167
Brehm, J. W. 138
Brinton, C. 12
Brown, P. A. 133
Brown, R. C. 10, 17, 33, 70, 76, 92, 97, 108, 143, 147, 150

Burgess, M. 133
Burke, K. 59
Burton, M. 58
Buss, A. M. 6, 55, 69, 134

Cantor, J. 42, 133, 148
Clark, R. A. 23
Cochran, S. 70
Cohen, C. E. 147
Cohen, M. E. 133
Converse, P. E. 73
Cronen, V. E. 58
Crotty, W. J. 14
Cutter, H. S. G. 111

Dabek, R. F. 149
DaGloria, J. 71, 76, 77, 92, 97, 136, 140, 143, 152
Davis, K. E. 15, 41, 59
Debus, G. 43
Deci, E. 41, 42, 45
Dengerink, H. A. 136, 137, 147
Denissen, K. 147, 148
DeRidder, R. 71, 76, 92, 97
Diener, E. 17
Dineen, J. 17
Dodge, K. 46
Dollard, J. 5, 8, 13, 55, 127
Donnerstein, E. 70
Doob, L. W. 5, 8, 13, 55, 127
Duda, D. 140
Duker, P. 147
Duncan, B. L. 149
Durkheim, E. 110
Dyck, R. J. 147

Ebbesen, E. B. 149
Eckensberger, L. H. 23, 29
Edelman, R. 133, 134
Edmunds, G. 36, 40
Elkind, D. 149
Embree, M. 70
Emminghaus, W. B. 23, 29
Endler, N. 47
Endresen, K. 17
Enzle, M. E. 41, 42, 45, 147
Epstein, S. 76

Faley, T. E. 13
Fanon, F. 14
Faucheux, C. 15
Feather, N. T. 23
Felson, R. B. 13, 92, 97, 108, 109, 111, 112, 115, 152, 153
Ferguson, T. J. 57, 76, 100, 145–149, 151–153
Feshbach, S. 7, 55, 124, 143, 152, 153
Fincham, F. 146, 147, 150
Fishbein, M. 144, 150
Forgas, J. P. 75
Fraser, S. C. 17
Frederiksen, N. 86
Freud, S. 5
Frodi, A. 41, 114, 133
Fuchs, R. 23

Gaes, G. G. 10, 108
Gambaro, S. 133
Garfinkel, H. 58
Geen, R. G. 134, 135
Geertz, H. 111

Gentry, W. D. 133, 135
Gergen, K. J. 53, 85, 152
Gibbs, J. P. 14
Goffman, E. 108
Goode, W. J. 110
Goodman, L. A. 118
Graumann, C. F. 72, 73
Green, J. A. 17
Greenwell, J. 136, 137, 147
Grice, H. P. 65

Habermas, J. 64
Hafner, A. 43, 46
Haner, C. F. 133
Hansen, R. 41
Harris, B. 72, 146, 147
Harris, M. B. 125
Hartshorne, H. 18
Harvey, J. H. 72
Harvey, M. D. 147, 148, 150
Heckhausen, H. 23–25
Heider, F. 7, 12, 58, 144, 146
Hepburn, J. R. 108
Herrmann, T. 71
Herzlich, C. 102
Hewitt, J. P. 111, 112, 133
Hewstone, M. 102
Hilke, R. 70
Hill, K. A. 149
Hoerster, N. 164
Hokanson, J. E. 133, 134
Hollingsworth, H. L. 143
Holsti, O. R. 11
Homans, G. L. 111
Horai, J. 13
Hume, D. 160

Janke, W. 43
Jaspars, J. 102, 146, 147, 150
Johnson, M. 66
Jones, E. E. 11, 15, 41
Joseph, J. M. 17, 76

Kagan, H.-J. 23
Kahn, M. 133
Kane, T. R. 17
Kaplan, A. 43, 46
Kaufmann, H. 55, 69, 70
Kelley, H. 41–43, 45, 47,
   145, 153
Kenny, A. 58
Kipnis, D. 14
Kirkham, J. S. 14
Konečni, V. J. 153
Kornadt, H.-J. 23, 25, 28, 29,
   37, 43, 46, 48, 103, 153

Koropsak, E. 134
Krauth, J. 81, 91
Kregarman, J. J. 136–137
Kulik, J. A. 147

Lacan, J. 59
Lagerspetz, K. M. 74
Lakoff, G. 66
Lalljee, M. 72
Lambert, W. W. 29
Lembert, E. M. 125
Levi-Strauss 59, 60
Levy, S. 14
Lienert, G. A. 81, 91
Lindskold, S. 13, 19, 71, 76,
   85
Linneweber, V. 27, 34, 35,
   66, 76, 77, 79, 80, 82, 83, 86,
   88, 89, 92–94, 96, 98, 99,
   101, 102, 143, 150, 152
Loeber, R. 22, 48
Löschper, G. 27, 34, 35, 66,
   76, 77, 79, 80, 82, 83, 86, 88,
   89, 92–94, 96, 98, 99, 101,
   102, 143, 150, 152
Lowe, C. 41
Lowell, E. L. 23
Lubek, I. 54, 127
Luckenbill, D. F. 108, 111
Luiten, A. 148, 149
Lyman, S. 59, 111
Lyons, D. 164

Macaulay, J. R. 17, 41, 114
Mackie, J. L. 169
Magnusson, D. 47
May, M. A. 18
McAra, M. 76
McClelland, D. C. 23, 30
McDonald, P. J. 85
Melburg, V. 10
Merz, F. 41
Michela, J. 41, 47
Mill, J. S. 161
Miller, N. E. 5, 8. 13, 55, 127
Miller, W. B. 111
Mills, C. W. 59
Misner, R. P. 14
Mixon, D. 18
Moos, R. H. 86
Moscovici, S. 15, 102
Mowrer, O. H. 5, 8, 13, 55,
   127
Mummendey, A. 15, 27, 28,
   33–35, 47, 57, 66, 74, 76,
   77, 79, 80, 82, 83, 86, 88, 89,

92–94, 96, 98, 99, 101, 102,
   143, 150, 152
Munzert, R. 71
Murray, H. A. 30, 46
Murstein, B. 46

Nacci, P. L. 17
Nagel, T. 102
Nesdale, A. R. 71, 76, 149,
   152
Newcomb, T. M. 73
Newtson, D. 75
Nickel, T. W. 76, 137, 147
Nozick, R. 162

O'Hagen, F. 36, 40
Ohbuchi, K. 36, 47
Olthof, T. 148, 149
Olweus, D. 22, 48
O'Neal, E. C. 85
Ossorio, P. G. 58

Palamarek, D. L. 148
Pastore, N. A. 136, 147
Pearce, W. B. 58
Pepitone, A. 21, 22, 29, 143,
   149
Perelman, C. 165
Peters, R. S. 58
Pettit, P. 72
Piaget, J. 35
Pisano, R. 135
Pittmann, T. S. 11
Price, R. H. 86
Psathas, G. 58
Putnam, H. 64

Quine, W. V. O. 51

Rabine, A. K. 133
Rawls, J. A. 162
Reitan, H. T. 76
Rescher, N. 168
Ribner, S. 108
Riess, M. 8, 100
Rivera, A. N. 10, 108
Roehl, C. A. 133
Rommetveit, R. 52, 57, 60
Rosenfeld, D. 149
Rosenfeld, P. 10
Rothmund, H. 41
Rubin, J. Z. 108
Rule, B. G. 57, 71, 76, 100,
   133, 145–153
Ryle, G. 7, 58

Sachs, L. 44
Schelling, T.C. 111
Schlenker, B.R. 13, 18, 19
Scott, M.B. 59, 111
Searle, J.R. 51, 65
Sears, R.R. 5, 8, 13, 55, 127, 152
Shantz, C.U. 149
Shaw, M.E. 76, 147
Shetler, S. 133
Shotter, J. 58
Siegel, M. 108
Smedslund, J. 58
Smith, R.B. 10, 33, 70, 76, 92, 97, 108, 143, 147, 150
Snyder, M. 149
Spiegel, S.B. 133
Stapleton, R.E. 17, 76
Steadman, H.S. 111, 115
Stephan, W.G. 149
Stokes, R. 111, 112

Stokols, D. 85
Stone, L. 133, 134
Stroebe, W. 101
Sulzer, J.L. 147

Tajfel, H. 85, 102, 101
Tanke, E.D. 149
Taylor, C. 60
Taylor, S.P. 17, 76, 133, 135
Tedeschi, J.T. 8, 10, 12, 13, 16–19, 33, 57, 70–72, 76, 85, 92, 97, 100, 108, 143, 147, 150, 152, 153
Thome, P. 41, 114
Thorndike, E.L. 11
Toch, H.H. 14
Todd, M.J. 59
Topitsch, E. 161
Turner, J.C. 102
Turner, R.H. 73

Van Roozendaal, J. 149

Walters, R.H. 55
Weiner, B. 23
Weinstein, E.A. 85
Weisfeld, G. 147
Werbik, H. 69, 71
Westman, M. 74
Willers, K.R. 134
Wittgenstein, L. 59
Worchel, P. 136–138

Zelin, M. 133
Zillmann, D. 42, 55, 69, 74, 133, 148
Zuckerman, M. 98
Zumkley, H. 37, 39, 42, 44, 46, 101, 103, 153

Subject Index

Accounts, 62, 111, 112, 118, 121, 124
Activation, 24, 25, 37, 38, 42, 43, 45
Actor-victim divergence, 92, 94, 95-98, 100, 103
Adequacy, 104
Affect, 27, 30
Aggression
  actor of, 4, 7–9, 11, 13, 14, 16, 33–35, 47, 55, 61, 65, 70–76, 78, 81, 82, 84, 92–101, 103, 121, 144, 145, 147–152
  approaches for research on, 33, 71, 100, 107, 108
  behaviorism and, 6
  characteristics of, 70
  concept of, 5, 8, 55, 57, 71, 75, 100, 157–159
  definition of, 5–10, 17, 19, 21, 107, 152
    attributional, 7–9, 16
    behavioristic, 5–7, 16
  definition criteria of, 34, 69, 71, 75, 76, 78, 81, 84
  disposition of, 23
  hostile, 16, 28, 36, 40, 41
  impulsive, 21, 22, 25, 30
  individual differences in, 153
  inhibition, 21, 25, 28, 29
  instrumental, 16, 21, 22, 28
  laboratory research on, 17, 18
  modeling and, 9

motive, 23–25, 28
observer of, 6, 7, 9, 16–18, 27, 41, 70, 72, 75, 97, 101, 103, 109, 144, 147, 148
perceived, 16, 17, 144, 146, 150
reactive, 35
  forms of, 22
as a set of behaviors, 6
spontaneous, 22, 25, 29, 35, 37, 40
theory, 21, 22, 152, 157
a versively stimulated, 101
victim of, 2, 13, 17, 27, 33–36, 38, 61, 70–72, 74–76, 79, 81, 84, 85, 92–99, 101, 103, 104, 110, 111, 131, 132, 134–140, 143, 144, 148, 151, 152
Aggressiveness, 21, 29, 36, 37, 40, 41, 46, 47, 149, 153
Aggressor, 27, 28, 36, 38, 61, 70, 71, 93, 103, 104, 111, 121, 123, 159, 160, 169
Alternatives of action, 70
  behavioral, 100
Anger, 12, 21–23, 28, 29, 37, 38, 40, 42–47, 62, 118–121, 124, 153, 168
Approach, 23, 24, 33–35, 48
Appropriateness, 34, 35, 72, 74, 84–86, 94–101, 103, 145, 147

Arousal
  autonomic, 133
  physiological, 41, 43, 56, 57
Attack, 13, 22, 37–43, 45, 46, 51, 55, 93, 99, 107–113, 115–118, 120, 121, 123, 125, 132–137, 148, 169
Attribution, 6, 10, 15, 16, 19, 36, 41, 42, 45–48, 76, 85, 98, 102, 131, 151
  causal, 4, 26, 43
Aversive
  condition, 22, 70
  consequence, 69, 70, 72, 75, 148
  event, 70, 101, 132, 137, 140
  state, 70
  stimuli, 69, 73, 100, 101
Avoidability, 100, 144
  of aversive events, 70
Avoidance, 23, 24

Behavior
  aggressive, 6, 24, 25, 33–36, 43, 47, 69, 70, 73, 76, 85, 102, 127, 128, 131, 132, 134, 135, 137, 138, 140, 158, 159
  motivated, 23
  setting, 85
Behaviorism and aggression, 5–7, 9
Bias, 47, 116, 148
Buss, Berkowitz aggression paradigm., 17, 70

Carelessness and forseeability, 145, 146
Catharsis, 25, 37, 38
Causal
  ambiguity, 41
  beliefs, 46, 48
  schemata, 46–48
Causality, 10, 144
Coercive
  influence, 72
  power, 16, 71, 72, 108, 109, 112, 121–125, 152
  theory, 19
Compensation, 81, 99
  by the harm-doer, 74
Conflict, 33, 75, 93, 108, 109, 111, 112, 116, 118, 119, 121, 124, 161–163, 165, 166, 170
Confrontation, 36–40, 75
Context, 51, 62–64, 66, 74, 75, 84, 85, 87, 92, 93, 102, 131, 165
  normative, 143
  situational, 72
  social, 14, 16, 33, 34, 58, 75, 87, 101–103, 151
Contract, 138, 139
Credibility, 13, 45, 109

Deactivation, 25, 37–39, 45, 46
Deprivation
  of expected gains, 12
  of resources, 12
Deterrence, 14, 110, 111
Development, 21–24, 28, 29, 35, 153
Discounting principle, 41, 42, 45

Ecological validity, 17, 18, 85
Electric shock, 69, 134, 139
Emotion, 1, 19, 25, 26, 28, 29, 41, 58, 93, 143, 169
Environment
  control of interindividual, 128–132
  of physical, 128–132
  physical, 11, 85, 129, 131, 132
Episode
  aggressive, 76
  social, 75, 77, 78, 82, 87, 92, 99
  videotaped, 92

Equity, 109, 110
Escalation, 27, 34, 74, 98, 99, 118, 119
Evaluation, 2–4, 13, 15, 21, 26, 30, 33, 35, 36, 48, 72, 74–78, 81, 84, 93, 95, 97–99, 102, 104, 143
  divergencies of 98, 100
  moral, 147, 163
Excuses and harm-doing, 15, 16, 19
Expectance x value theory, 25
Expectancy, 23, 24, 149, 150

Factors
  biological, 21, 29
  situational, 22, 27, 112, 148
Failure, 12, 26, 132–135
Forseeability, 145, 146
Frustration, 16, 25, 29, 37, 76, 77, 127, 131–134, 137, 138
  aggression hypothesis, 5, 8, 13, 18, 21, 29, 33, 127, 131, 132
  arbitrary vs justified, 136–139
  expected vs unexpected, 136–138
Frustrative cues, 29

Goal
  aggressive, 25, 40
  non-destructive, 24
Guilt, 25, 36, 165

Harm, 2, 7, 8, 22, 33–35, 40, 61–66, 70–72, 77, 79, 81, 84, 99, 101, 103, 107, 115, 144, 165
  severity of, 148, 149
Harm-doing, 6–10, 13, 110
  accidental, 6, 15
  antinormative, 18
Hostility, 36, 39, 56, 57, 70, 138

Identity, 13
  face saving, 107–110, 112, 113, 121–125
Ideology, 103, 162
Impression management, 4, 11, 13, 16, 100, 107, 108, 112, 113, 121–125
  strategy, 11
Incentive, 24, 25, 29, 36

Individual differences, 22–24, 27–29, 34–36, 40, 41, 45, 47, 48, 103
  in aggressiveness, 36, 37, 40, 41, 46–48
  in attribution of intent, 35, 36, 47
  in causal schemata, 48
  in reactive aggression, 34
  in spontaneous aggression, 34
Inhibition, 24, 43, 44, 46
Injury, 24, 33, 38, 55, 70, 74–79, 81, 82, 84, 86, 89, 93, 101, 103, 146, 148
Insult, 28, 108, 111, 113, 115, 117, 118, 120, 121, 125, 133–135
Intent, 2, 7–9, 22–25, 28, 30, 34, 36, 38, 41, 47, 56, 61–66, 71, 72, 75–79, 81, 82, 84, 86, 89, 101, 136, 137, 158, 159
  aggression and, 7, 17
  ambiguous, 42, 46
  benign, 46
  to harm, 6, 56
  harmless, 42, 45
  hostile, 22, 35, 40–43, 46
Intentionality, 137, 144
Interaction, 24, 27, 28, 30, 33, 34, 42, 43, 72–76, 84–87, 89, 92–97, 100, 101, 152
  aggressive, 3, 4, 27, 28, 35, 36, 40, 41, 45–47, 93, 101, 112, 121, 125, 131, 158, 159, 162, 170
  course of, 75, 98, 99
  person-situation, 46, 47
  process, 20, 74
  sequence of, 4
  social, 3, 48, 71, 72, 75, 76, 103, 153, 161, 168
Interpretation, 13, 27–29, 47, 55, 57, 82, 84, 152, 153
  divergencies of, 74, 75

Jugdement, 33, 40, 41, 56, 74, 75, 78, 85–87, 89, 93, 94, 97, 99, 100, 107, 134, 140, 143, 147, 149
  moral, 15, 23, 35, 76, 100
  social, 55, 74
Justice, 61, 65, 103, 109, 111, 129, 136, 137, 143, 147, 158
  corrective, 164, 165

distributive, 110, 164–167
procedural, 161–164, 169, 170
restoration of, 103
substantive, 161–164
Justification, 8, 27, 28, 34, 77, 100, 111, 136, 138, 139, 144, 159–161
aggression and, 8, 9
harm-doing and, 13, 15–18
of hostile responses, 76

Language System, 58

Morality, 159–161
Moral obligation, 101
Motivation, 23–25, 30, 35–38, 40, 41, 46–48, 58, 158
theory, 23, 27–29
of aggression, 24, 28, 30
Motive, 23, 29, 30, 47, 48, 55–57, 103, 111, 132, 134, 137, 143–148, 152
to avoid aggression, 24
Myths, 103
Negotiation, 57, 64, 66
Neuralgic places, 85, 86
Nonviolent resistance, 15
Norm 2, 3, 27, 34–36, 40, 70, 71, 74, 76–78, 85, 93, 99, 129, 138, 139
(see also violation)
deviation, 33, 34, 36, 40, 76, 78, 79, 81, 84, 86
of negative reciprocity, 13, 18, 34, 35, 74
retaliation, 34–36, 40
situationally relevant, 17
Noxious stimulation, 12, 17

Ostensive definition, 51–53
Oughts, 143
causal responsibility and, 144

is – ought discrepancy, 145, 147, 148
normative beliefs and, 143–145, 147

Person perception, 10, 11, 72
Perspective, 2, 33, 47, 48, 70, 74, 75, 92–95, 98
of the victim, 34
Pragmatic consequence, 53
Processes
affective internal, 21
cognitive, 30
internal, 21
Progress of aggressive inter-action, 75
Provocation, 37, 40, 42, 72, 76, 77, 94, 147
Punishment, 10, 12, 13, 15–18, 25, 34, 36, 40, 61, 65, 66, 78, 84, 143, 145, 162–166, 169
attention getting and, 14
definition of, 11, 12
functions of, 13, 14
social, 13
control and, 107, 109–112, 121, 125
symbolic functions of, 14
types of, 11

Re-action, 35, 95, 98
Reciprocity, 35, 35
Responsibility, 42, 44, 45, 144
blame and, 147
levels of, 145, 146, 150, 151
Retaliation, 34–41, 61, 94, 108, 114, 123, 124, 143, 145, 147, 148, 165, 169
Retribution, 65, 110, 159, 164, 165, 169, 170
Rules
general, 27, 30
social, 2, 3, 54, 85, 93

Segment, 18, 75, 93–96, 99
Segmentation of action, 75
Self-evaluation, 132, 135, 137

Sequence of action, 24, 73, 75
Social situation, 25, 30, 69, 72, 84, 86
vacuum 102
Stereotyping, 149, 151
Structural nucleus, 61, 62, 66, 67
Subjective representations, 7

Taxonomy
behavior specific, 16, 87, 92, 102
of situations, 84–86
of threats and punisment, 15
Tendency, 34, 35, 58
enduring, 22, 27, 30
for completion, 71
'gute Gestalt', 71
Terrorism, 14
Threats, 10, 11, 13, 15, 16, 30, 108–111, 113, 115–121, 124, 125, 133
definition of, 11
perceived, 11
purposes of, 11

Unit
functional, 23
social, 73, 75
Unpacking
horizontal, 62
structural, 57, 59, 60, 62, 63, 65, 66
vertical, 62
Utilitarism, 168, 169

Values, 28, 162
Violation
norm, 2, 13, 34, 40, 71, 72, 75, 81, 102, 103, 111, 113, 120, 121, 123, 125, 144, 145
rule, 112, 114–125
Violence, 12, 65, 69, 70, 102, 108, 110, 111, 113, 118, 119, 149